CAMBRIDGE COUNTY GEOGRAPHIES

General Editor: F. H. H. GUILLEMARD, M.A., M.D.

SUSSEX

Cambridge County Geographies

SUSSEX

by

GEORGE F. BOSWORTH, F.R.G.S.

With Maps, Diagrams and Illustrations

Cambridge:

at the University Press

1909

CAMBRIDGE UNIVERSITY PRESS
Cambridge, New York, Melbourne, Madrid, Cape Town,
Singapore, São Paulo, Delhi, Mexico City

Cambridge University Press
The Edinburgh Building, Cambridge CB2 8RU, UK

Published in the United States of America by Cambridge University Press, New York

www.cambridge.org
Information on this title: www.cambridge.org/9781107646339

© Cambridge University Press 1909

First published 1909
First paperback edition 2013

A catalogue record for this publication is available from the British Library

ISBN 978-1-107-64633-9 Paperback

CONTENTS

PAGE

1. The South-Eastern Peninsula. County and Shire.
   The Word *Sussex*. Its Origin and Meaning .   1

2. General Characteristics. Position and Natural Conditions. Scenery 5

3. Size. Shape. Boundaries 9

4. Surface and General Features 13

5. Watershed. Rivers 17

6. Geology and Soil 20

7. Natural History 28

8. Climate and Rainfall 33

9. The Coast—Gains and Losses. Its Protection—Sea
   Walls and Groynes. Lighthouses and Lightships . 38

10. People—Race, Dialect, Settlements, Population . 45

11. Agriculture. Main Cultivations, Woodlands, Stock . 50

12. Industries and Manufactures 55

13. Minerals. Exhausted Mining Industries . . 60

14. Fisheries and Fishing Stations 67

15. Shipping and Trade—The Chief Ports. Extinct Ports.
    Cinque Ports 71

PAGE

16. History of Sussex 76

17. Antiquities—Prehistoric. Roman. Saxon . . 84

18. Architecture. (a) Ecclesiastical—Churches, Cathedral,
     Abbeys 91

19. Architecture. (b) Military—Castles . . . 100

20. Architecture. (c) Domestic—Famous Seats, Manor
     Houses, Cottages 105

21. Communications—Past and Present—Roads, Rail-
     ways, Canals 111

22. Administration and Divisions—Ancient and Modern 117

23. The Roll of Honour of the County . . . 123

24. The Chief Towns and Villages of Sussex . . 129

ILLUSTRATIONS

PAGE

Chichester from the Canal	3
The Causeway, Horsham	7
The Downs near Wannock	8
Rye	11
Wytch Cross	14
Chanctonbury Ring	16
The Arun at Arundel	18
Fittleworth Bridge	20
View from the Devil's Dyke	27
Sun Oak, St Leonard's Forest	29
Arundel Castle	31
Eastbourne	38
Worthing Sands	41
Brighton from the West Pier	42
Market Cross and Cathedral, Chichester	46
Strand Gate, Winchelsea	48
Oxen at Work on a Sussex Farm	53
Shoreham and the River Adur	56
Bosham	57
Pulborough Church	61
Hurstmonceux Castle	63
Hammerpond Waterfall	66
The Dieppe Boat leaving Newhaven	69
Beachy Head	70
Ypres Tower, Rye	73
Hastings Castle: the Chancel Arch of the Chapel	75
The Gateway, Battle Abbey	79
Battle Abbey: the Spot where Harold fell	82
Palaeolithic Implement	86
Neolithic Celt of Greenstone	86
The Long Man, Wilmington	89

ILLUSTRATIONS

PAGE

East Lavant Church 90
Malling Hill, Lewes 92
The Porch, Bishopstone Church 94
New Shoreham Church 96
Worth Church 97
Camber Castle 100
Pevensey Castle 102
Bodiam Castle 104
The Great Hall, Mayfield 107
Brede Place 108
Cowdray House, Midhurst 109
Old Houses, Petworth 110
Chichester Cathedral 112
Lewes Castle: the Entrance Gate . . . 115
Battle Abbey: the Cloister Front . . . 120
Christ's Hospital, Horsham 122
Parsonage Hall, West Tarring . . . 124
Edward Gibbon 126
Field Place, Warnham (Shelley's Birthplace) . . 127
Petworth Church 135
Farm House, Warnham 138
Diagrams 140

MAPS

England and Wales, showing annual rainfall . . . 35

The Illustrations on pp. 16, 27, 31, 38, 42, 57, 70, 79, 97, and 112 are from photographs by Messrs F. Frith & Co., Ltd., Reigate; those on pp. 8, 29, 48, 56, 66, 69, 73, 89, 94, 100, 127, and 138, are from photographs by the Homeland Association, Ltd., London; those on pp. 75, 82, 102, 115, and 120 are from photographs by Mr A. P. Wire, Leytonstone; and that on p. 122 is from a photograph by M. Buê, Christ's Hospital.

1. The South-Eastern Peninsula. County and Shire. The Word *Sussex*. Its Origin and Meaning.

The three counties of Kent, Surrey, and Sussex form a compact peninsula in the south-east of England, having the Thames on the north, and the sea on the east and south, while the boundary on the west is formed by Berkshire and Hampshire. This south-eastern peninsula has always been of the greatest importance in our history, for the chief lines of communication between the Continent and London pass through one or other of the three counties. The Thames is the waterway to London; the roads from Dover, Hastings, Brighton, and Portsmouth are the highways through this peninsula to the metropolis; and the railways from the chief seaports of the south-east carry passengers and goods to the great city.

It will thus be readily understood that each of these three south-eastern counties is of considerable importance on account of the proximity of London; and it is both interesting and instructive to have a definite knowledge of the past and present condition of all of them. In this

book we are concerned only with Sussex, one of the
" home counties " as it is called ; and it will be well at
the outset to discover what is meant by a county, and
then find out how Sussex came by its name.

If we look at a map of England we notice that some
of the counties are large and some are small ; and we also
find that some of the names end in *-shire*. Why is there
this difference in size and name ? It is said in some
books that King Alfred divided England into counties.
This, of course, is quite wrong, for although that great
King did many things in his eventful reign, he certainly
did not bring about the division of England into counties.
We know that while some of the counties existed before
his time, others were not formed till long after his death.

The fact is that some of our present counties, such as
Kent, Surrey, and Sussex, are survivals of the old English
kingdoms, which have kept their former names and extent.
Others of our counties are *shires*, or shares of former large
kingdoms, such as Mercia, or Wessex, or Northumbria.
Thus Staffordshire was once a part of Mercia, Yorkshire
of Northumbria, and Hampshire of Wessex. It may be
said quite correctly that our English counties have grown,
and it is this gradual growth that makes their history so
interesting.

When we investigate the boundaries and extent of
the counties we find that they often represent the districts
of tribes or kingdoms. Thus Kent was first the possession
of a British tribe, the Cantii, which was afterwards conquered by the Jutes ; and Sussex was a kingdom formed
by the Saxons in the fifth century. The present county

of Sussex corresponds more or less to that ancient kingdom, though it may at one time have extended further to the west.   The story of the colonisation of Sussex is given in the *English Chronicle*, and we shall make further reference to this fact in another chapter.   Here we may notice in

Chichester from the Canal

passing that Aelle is generally recognised as the founder and first king of Sussex, and that he landed near the present city of Chichester in 477 and did not complete his work of conquest till 491.   For a period of upwards of fourteen centuries Sussex has ranked as one of the

English kingdoms, or counties, and of all the English counties it is the most typical, and, perhaps, the most natural.

This is a very remarkable fact, and one of the deepest interest to us in our study of the geography and history of Sussex. We shall understand these much better if we grasp this fact clearly, that many of our counties are the same, or nearly the same, as the first English kingdoms, which were never less than seven in number, and often far more numerous. If we look at the map of England it will be seen that the physical features of Sussex mark it out at once as a distinct and separate region ; and its history shows it as always an independent kingdom, or a well-defined county, keeping the same essential boundaries throughout its entire existence. Even when Wessex conquered Sussex, the kingdom of the South Saxons continued to have its own under-kings. When Sussex gradually dropped from the rank of a kingdom to that of a county, it came to be amalgamated with the rest of England.

There is thus no difficulty in tracing the origin of its name, which it obtained from the Saxon conqueror. Essex the land of the East Saxons, Wessex the land of the West Saxons, and Sussex the land of the South Saxons are all quite obvious in their origin. The Saxons were a Teutonic people who first began to trouble the British coasts before the Romans went away. They came from the districts we now call Holstein, Westphalia, Hanover, and Brunswick ; and wherever they settled they called the land after their own name. This is no doubt the reason why the Keltic people in Scotland, Wales, and Ireland call all Englishmen *Saxons* to this day; but we English people must remember that *Saxon* by itself always meant the people of those parts only where the Saxons settled.

It is now quite clear why Sussex is a separate county, and why its boundaries should be what they are. We may look upon Sussex as a typical instance of an old English kingdom becoming a county and retaining a certain local independence of its own to the present day.

2. General Characteristics. Position and Natural Conditions. Scenery.

In the previous chapter we have seen that Sussex is perhaps the most typical, and the most natural, of all the English counties. Its physical features mark it out as a distinct and separate whole, and its history shows that it has always been either a well-defined kingdom or county, preserving the same boundaries and extent throughout its existence. If we study a good map of the county, we find a long spur of chalk, forming the South Downs, runs through it like a backbone till it terminates at Beachy Head.

Between the South Downs and the coast there is a narrow belt of lowland, and this belt, small as it is, really comprises the whole of historical Sussex. On their northern side, the Downs descend by a steep escarpment into the wide valley of the Weald, of which a broad view is gained from the summit of the Devil's Dyke, near Brighton. The country between the North and South

Downs was once covered by chalk, but this has been worn away, and the district is now occupied by the soft, muddy Weald clay, and the harder beds of Hastings Sand. It will be seen that this wide tract extends along the northern border of the county from the Downs to the boundaries of Kent and Surrey, and from Petersfield in Hampshire to Pevensey, Hastings, and the Romney Marshes.

The Sussex Weald was for many ages untilled and uncleared, and formed a great stretch of forest known to the Romans as *Silva Anderida*, and to the English as the *Andredesweald*. The cold clay of the Weald can support little more than trees, and even in our own days it is only scantily cultivated. In early times, this belt of forest was dense and trackless, forming a barrier to intercourse with other parts of the country; and it is this isolation of Sussex by the Weald and the Marshes which makes the history of Sussex so peculiar and so typical.

It will be seen in a later chapter that Sussex is essentially an agricultural county, and as such enjoys many advantages, owing largely to its position and climate. It has a long sea coast fronting the English Channel from the borders of Hampshire to some distance beyond Rye; and its climate, bracing and healthy in the higher parts, is genial and salubrious near the sea. From Bognor in the west to Hastings in the east there is a succession of delightful watering-places—holiday and health resorts which are thronged in the season by thousands of people from our great crowded cities. The influx of these seekers after pleasure and health makes Sussex familiar to all sorts and conditions of men, and brings some amount of prosperity to a county that has suffered from agricultural depression and the fall in prices. The number of large towns in Sussex is restricted, and the greater part of the population is gathered in the towns on the coast. The district of the Weald has only two small

The Causeway, Horsham

towns, Horsham and Midhurst; and its villages support a scanty population by agriculture, hop-picking, dairying, and cattle-rearing.

Sussex will always be associated with the chief event in our nation's story, for it was at Pevensey, in 1066, that William the Norman landed, and having defeated his enemy at Hastings, proceeded to make himself the master of our country. The battle of Hastings is the great turning-point that marks the close of one epoch, and the beginning of a new era. There are some who assert that Julius Caesar landed in Sussex, but there is far more evidence that the Romans entered England by Kent.

The Downs near Wannock

The landing of William at Pevensey was probably a far greater event than the coming of Julius Caesar, and Sussex people may well be content to remember that it was one of the first English kingdoms to be formed by Aelle, one of the first to be Christianised by Wilfrid, and the scene of the contest that made England the kingdom of William the Conqueror.

The scenery of Sussex does not vie with that of Wales, or Derby, or the Lake district, but it has a type of natural beauty that gives it a peculiar charm. It has neither rivers of striking beauty, nor mountains of any great height. It has no lakes, and its sea-coast scenery is not so fine as that of Yorkshire or Devonshire. Yet the scenery of the South Downs and the Forest Ridge is exceeded in beauty and interest by few parts of England. It has been well said that "the crown of the county's scenery is the Downs, and the most fascinating districts are those which the Downs dominate."

Perhaps no modern writer has done more to extol Sussex than Rudyard Kipling, who is a resident in the county. Here is a stanza from one of his Sussex poems:

> "God gave all men all earth to love,
>   But since man's heart is small,
> Ordains for each one spot shall prove
>   Beloved over all.
> Each to his choice, and I rejoice
>   The lot has fallen to me
> In a fair ground—in a fair ground—
>   Yea, Sussex by the sea!"

3. Size. Shape. Boundaries.

In the two previous chapters we have learnt how Sussex, once a Saxon kingdom, came to be a county, and we have read about its characteristics. We are now, with the aid of the map, going to study its size, shape, and boundaries.

First with regard to the size of the county, we may say that the present county has varied little since it was first formed more than a thousand years ago. Perhaps it then extended a little more to the west, but on the whole we shall be tolerably safe in saying that its area has been much the same in the whole course of its history. We are of course speaking of the ancient county, for there is also another way in which some of the counties are named since the passing of the County Councils Act in 1888. It was then arranged that a few of our English counties should be subdivided for purposes of administration, and under this Act the ancient county of Sussex was divided into the administrative counties of West Sussex and East Sussex. This will make it quite clear that whenever we refer to the larger Sussex we shall speak of it as the Ancient County, and when we refer to the two divisions we shall speak of them as the Administrative County of East Sussex, and the Administrative County of West Sussex.

Sussex is a maritime county in the south-east of England, lying along the English Channel. It has an area of 933,887 acres, or about 1459 square miles. It is thus somewhat smaller than Kent, nearly twice as large as Surrey, and occupies about one thirty-fourth of the entire area of England and Wales. Of the two divisions, the Administrative County of East Sussex is the larger, having an area of 528,807 acres, while West Sussex has an extent of 403,602 acres. A small portion of Sussex is included in the adjacent counties of Kent, Surrey, and Hampshire for local government purposes.

The greatest measurement of Sussex from east to west is about 73 miles, and 27 miles is its extreme breadth from north to south. The average of the latter is probably about 20 miles, but it will be noticed that the county narrows considerably in the east. The most northerly point is where the Oxted and Groombridge railway line leaves

Rye

the county, while Selsey Bill is the most southerly point. The extreme point on the east is about five miles east of Rye, and on the west the extreme point is just to the north of Rowland's Castle.

Sussex is one of the most compact of the English counties, being in shape somewhat of an irregular parallelogram. Its boundaries were settled long before the Norman conquest, and it is not at all difficult to trace their origin. The English Channel from east to west is the boundary on the south side, and there is every reason to believe that there were two settlements of the Saxons, one at Chichester and another at or near Pevensey. Thus it came about that the Saxon settlers took the whole coast-line from beyond Pevensey to Chichester. Having established themselves along the sea-coast the invaders slowly worked their way northwards. This was a long and difficult affair, for Sussex was then covered with the great and dense forest of Andredsweald. At length, however, the South Saxons pushed their way till they reached a high ridge in the midst of the district we now call the Weald. There they found the Saxons from the north were in possession of the land north of this ridge, and as the outcome of some agreement there is no doubt that this dividing line was adopted as the boundary between Sussex on the south and Surrey on the north.

On the west, Hampshire forms the boundary, which was drawn through the extensive woodlands on this side. The boundary on the east and north-east is artificial and irregular, and divides the county from Kent. It is not necessary to go into details, but it is generally thought that the Kentish kingdom had pushed itself as far west as it wished to go, and beyond that line Sussex was allowed to exist. We can now see that Sussex was originally a kingdom between Wessex on the west and Kent on the east, and between Surrey on the north and

the English Channel on the south.   Its physical features
mark it out at once as a distinct and separate whole ; and
its history shows it as always an independent kingdom
or a well-defined county, preserving the same boundaries
throughout its entire existence.

4. Surface and General Features.

A glance at the map will show us that so far as
regards the surface of the land Kent, Surrey, and Sussex
have many points in common.   Each county is intersected
by a range of hills that runs from west to east, and each
has within its borders the remains of the ancient forest
formerly known as Andredsweald.   The physical features
of Sussex are somewhat similar to those of Kent, but they
have a more marked character, and the rugged south-
eastern portion of the county resembles the hilly and
beautiful region of south-west Surrey.

The physical features of the county may be considered
under three divisions.   First, there is the Weald of Sussex
in the northern part ; secondly the high lands, known
as the South Downs, and the Wealden Heights or Forest
Ridge in the south-east ; and thirdly the low coastal plain
stretching from Brighton to Selsey.

The Weald was till modern times one of the wildest
and least inhabited parts of England.   The heights
which form the water-parting were covered with dense
masses of forest, and were traversed only by narrow
lanes or mule-tracks.   The country is now brought under
cultivation, but is still thinly peopled, and has, as we

have already seen, only two small towns, Horsham and Midhurst. The Weald of Sussex extends from Pevensey Bay to the hills beyond Petworth, and, although called a plain, it has several hills of 300 feet and over.

Wytch Cross

The Forest Ridge is the highest part of the Hastings Sand, and has some very picturesque scenery, quite different from that of the other natural divisions of the county. It is exceeded in beauty and interest by few parts of England.

The northern portion begins east of Horsham and reaches a height of about 500 feet. Its western part is thickly wooded by the forests of St Leonard, Tilgate, and Worth; but in the east the woods are not so noticeable. At Wytch Cross the hills reach a height of 658 feet, and then follow the wild uplands of Ashdown Forest, one of the most beautiful parts of Sussex. The Forest Ridge rises to its highest point at Crowborough Beacon, which is 792 feet above the sea. The southern portion of the Forest Ridge may be distinctly traced as the southern boundary of the Rother valley, and becomes a marked line of hills passing by Heathfield till it attains the height of 620 feet at Brightling Beacon. The Forest Ridge continues in a south-easterly direction till it reaches the sea at Fairlight, near Hastings.

The most remarkable feature in the surface and scenery of Sussex is the bold and open range of chalk hills, called the South Downs, extending into it from Hampshire, and stretching in nearly an easterly direction for a distance of 53 miles to Beachy Head, which rises perpendicularly above the shore to a height of over 500 feet. Their northern declivity is precipitous, but on the south their descent is gradual, except in the neighbourhood of Brighton, where they form some lofty cliffs. Gilbert White described the South Downs as a "chain of majestic mountains." This of course was the exaggeration of a home-keeping man who truly loved this district. The average height of the South Downs is about 500 feet, and the highest points are Duncton Down, 837 feet, Linch Down, 818 feet, and Ditchling Beacon, 813 feet.

The views from the Downs are exceedingly fine and
extend far over the Weald on the north, or towards
the sea-coast on the south. Picturesque villages nestle
under the foot of the hills, and the clusters of cottages,
some thatched, some tiled, and all built of flint or boulders,
are quite characteristic of Sussex. The South Down
shepherds are now almost extinct. They were a very

Chanctonbury Ring

peculiar race, and used to live in caves, or in huts dug
into the side of a bank, and lined with heath or straw.

The coastal plain, southward of the Downs, extends
from their base to the sea. It is a fertile and well-
cultivated district, which, in its eastern extremity between
Brighton and Shoreham, is, for the most part, about a mile
in width. Between the rivers Adur and Arun the plain

increases to three miles, and from the Arun westward to the borders of Hampshire its breadth varies from three to seven miles. Extensive tracts of marshland are adjacent to the coast between Pevensey and Rye, and others are situated on the lower parts of the course of the rivers Ouse, Adur, and Arun.

5. Watershed. Rivers.

The river system and drainage of Sussex belong to a region that stretches from Wiltshire on the west to Kent on the east, and the map will show that the watershed of this portion of England is along the line of the Wealden uplift. From this high land the Wey, the Mole, and the Medway flow northwards into the Thames, and the Salisbury Avon, the Itchen, and, in Sussex, the Arun, and the Sussex Ouse flow southwards to the sea. It is also worth noticing that all the more eastern rivers of this region rise in the centre of the Weald, and cut gaps through the North or South Downs. In Sussex, for instance, the Arun and the Ouse flow through gaps near Arundel and Lewes, and the same may be said of the small Sussex rivers, the Adur and the Cuckmere.

The Sussex rivers are tolerably numerous, but are all of small size, and of little commercial importance. They all rise within the county, and, with the exception of the Mole, which rises in the Forest Ridge and enters Surrey, every stream has the same direction from the north to the English Channel. Commencing on the west, the chief rivers are the Arun, the Adur, the Ouse, the Cuckmere, the Ashburne, and the Rother.

The Arun has its main source in St Leonard's Forest, but some of its feeders pass west of the Forest Ridge and

The Arun at Arundel

flow from further northwards. The main stream passes Horsham and after a short distance turns south near Slinfold. When it reaches Stopham its waters are increased by a tributary, the West Rother, which rises in the north-west of the county. Its course is now very winding and it flows through a rich tract of marshes, and by the town of Arundel to the sea at Littlehampton. The tide flows up the Arun for a distance of seventeen miles to Amberley. The scenery of the Arun is generally tame, but its broad stream adds to the charm of the district around Arundel. The river is celebrated for its mullets.

The Adur also rises in St Leonard's Forest, and pursues a southward course by Steyning and Bramber to Shoreham, where it takes an easterly direction, nearly parallel with the coast, falling into the sea a little to the west of Brighton.

The Ouse is formed by the junction of two streams, one of which rises in the forest of Worth, and the other in that of St Leonard. They unite near Cuck-field, whence the Ouse, proceeding first eastward, and then southward, passes the town of Lewes to the sea at Newhaven.

The Cuckmere, a very small river, rises near Heath-field, and flows into the sea at Cuckmere Haven, near Beachy Head.

The Rother has its source at Rotherfield, near Ash-down Forest, whence it proceeds eastward and soon becomes the boundary between Sussex and Kent. After passing the Isle of Oxney, in the latter county, it suddenly turns southward across the eastern extremity of Sussex. It then expands into an estuary, and reaches the sea below the town of Rye, whose harbour it forms. The mouth of the Rother was formerly at New Romney, but in the reign of Edward I it deserted its old channel for the present one.

Fittleworth Bridge

6. Geology and Soil.

By Geology we mean the study of the rocks, and we must at the outset explain that the term *rock* is used by the geologist without any reference to the hardness or compactness of the material to which the name is applied; thus he speaks of loose sand as a rock equally with a hard substance like granite.

Rocks are of two kinds, (1) those laid down mostly under water, (2) those due to the action of fire.

The first kind may be compared to sheets of paper one over the other. These sheets are called *beds*, and such beds are usually formed of sand (often containing pebbles), mud or clay, and limestone, or mixtures of these materials. They are laid down as flat or nearly flat sheets, but may afterwards be tilted as the result of movement of the earth's crust, just as you may tilt sheets of paper, folding them into arches and troughs, by pressing them at either end. Again, we may find the tops of the folds so produced washed away as the result of the wearing action of rivers, glaciers and sea-waves upon them, as you might cut off the tops of the folds of the paper with a pair of shears. This has happened with the ancient beds forming parts of the earth's crust, and we therefore often find them tilted, with the upper parts removed.

The other kinds of rocks are known as igneous rocks, which have been melted under the action of heat and become solid on cooling. When in the molten state they have been poured out at the surface as the lava of volcanoes, or have been forced into other rocks and cooled in the cracks and other places of weakness. Much material is also thrown out of volcanoes as volcanic ash and dust, and is piled up on the sides of the volcano. Such ashy material may be arranged in beds, so that it partakes to some extent of the qualities of the two great rock groups.

The production of beds is of great importance to geologists, for by means of these beds we can classify the rocks according to age. If we take two sheets of paper, and lay one on the top of the other on a table, the upper one has been laid down after the other. Similarly with two beds, the upper is also the newer, and the newer will remain on the top after earth-movements, save in very exceptional cases which need not be regarded by us here, and for general purposes we may regard any bed or set of beds resting on any other in our own country as being the newer bed or set.

The movements which affect beds may occur at different times. One set of beds may be laid down flat, then thrown into folds by movement, the tops of the beds worn off, and another set of beds laid down upon the worn surface of the older beds, the edges of which will abut against the oldest of the new set of flatly deposited beds, which latter may in turn undergo disturbance and renewal of their upper portions.

Again, after the formation of the beds many changes may occur in them. They may become hardened, pebble-beds being changed into conglomerates, sands into sandstones, muds and clays into mudstones and shales, soft deposits of lime into limestone, and loose volcanic ashes into exceedingly hard rocks. They may also become cracked, and the cracks are often very regular, running in two directions at right angles one to the other. Such cracks are known as *joints*, and the joints are very important in affecting the physical geography of a district. Then, as the result of great pressure applied sideways, the rocks may be so changed that they can be split into thin slabs, which usually, though not necessarily, split along planes standing at high angles to the horizontal. Rocks affected in this way are known as *slates*.

If we could flatten out all the beds of England, and arrange them one over the other and bore a shaft through them, we should see them on the sides of the shaft, the newest appearing at the top and the oldest at the bottom, as shown in the figure. Such a shaft would have a depth of between 10,000 and 20,000 feet. The strata beds are divided into three great groups called Primary or Palaeozoic, Secondary or Mesozoic, and Tertiary or Cainozoic, and the lowest of the Primary rocks are the oldest rocks of Britain, which form as it were the foundation stones on which the other rocks rest. These may be spoken of as the Precambrian rocks. The three great groups are divided into minor divisions known as systems. The names of these systems are arranged in order in the figure with a very rough indication of their relative importance, though the divisions above the Eocene are made too thick, as otherwise they would hardly show in the figure. On the right hand side, the general characters of the rocks of each system are stated.

With these preliminary remarks we may now proceed to a brief account of the geology of the county.

The main geological divisions of Sussex belong to what is called the Valley of the Weald, and are largely connected with the history of the chalk formation. Most of the formations belong to the group called Secondary, of which the chalk is uppermost; but in the west of Sussex there is a flat stretch formed by Tertiary Strata. The geological map will help us to realise that the central part of the Weald is formed by the Hastings Sands, round which there lies in a ring the Wealden Clay. Then

	Names of Systems		Characters of Rocks
TERTIARY	Recent & Pleistocene Pliocene Eocene		sands, superficial deposits

clays and sands chiefly |
| SECONDARY | Cretaceous | | chalk at top
sandstones, mud and clays below |
| | Jurassic | | shales, sandstones and
oolitic limestones |
| | Triassic | | red sandstones and marls, gypsum and salt |
| PRIMARY | Permian | | red sandstones & magnesian limestone |
| | Carboniferous | | sandstones, shales and coals at top
sandstones in middle
limestone and shales below |
| | Devonian | | red sandstones,
shales, slates and limestones |
| | Silurian | | sandstones and shales
thin limestones |
| | Ordovician | | shales, slates,
sandstones and
thin limestones |
| | Cambrian | | slates and
sandstones |
| | Pre-Cambrian | | sandstones,
slates and
volcanic rocks |

succeeds the ring of the Lower Greensand, which is encircled by the Gault and Upper Greensand; and lastly there is the broad belt of the Chalk.

The oldest strata in Sussex come to the surface in the Weald, although some older deposits were penetrated by a deep boring near Battle. It may be stated generally that while the newest deposits occur in the south and south-west, the oldest rocks are found in the north-east. These belong to the Jurassic system, and although attempts have been made to reach the Coal Measures, they have up to the present been without success.

The Cretaceous rocks follow the Jurassic without a break, and in their lower part the deposits consist mainly of sands, commonly known as Hastings and Tunbridge Wells Sands. Here it may be mentioned that when some borings were recently made at Waldron, an inflammable gas was struck in the Fairlight Clay. This gas was found to be a genuine petroleum, and has been used for lighting the railway station and offices.

The Lower Greensand succeeds these last strata and is mainly sandy deposits with beds of harder rock. The sands are commonly striped with *green*, from the presence of small grains of a dark green mineral—glauconite. This mineral is an iron compound, which readily oxidises when exposed, and then the sands take the buff or rusty hue which makes people wonder why they are called *Green*sand.

Above the Lower Greensand is the Gault, a stiff, dark blue clay about 300 feet thick and fossiliferous. The Upper Greensand follows, and is from 90 to 100 feet thick in West Sussex. The Greensand passes into the Chalk, which covers nearly one-third of Sussex, and forms a sharply defined region unlike any other in the county, known as the South Downs. The Chalk has a thickness of about 1000 feet in Sussex, and forms a dry region with no springs or flowing water, except in the lowest valleys. The greater part of the Downs forms open, rolling country, bare and treeless, but covered with excellent pasture, or with light calcareous soil readily worked by the plough.

The Chalk formation may be considered under three divisions: first, the Lower Chalk, which consists of greyish marl in alternate hard and soft beds which make conspicuous ledges in the foreshore, and at the base of cliffs between Eastbourne and Beachy Head. The Lower Chalk has a thickness of 150 to 200 feet ; and although not much water is obtained from it, hydraulic lime is produced in considerable quantities. Second, the Middle Chalk, which is hard, splintery, and full of fossils ; and thirdly, the Upper Chalk, which is purer white, and softer than the Middle, and yielding flints, which are almost confined to it. The best places to study the Chalk in Sussex are the cliffs between Eastbourne and Brighton, and in the large pits near Lewes.

The Chalk is succeeded by the Woolwich and Reading series of rocks. Here we find quite a change of conditions, for we are now in the Eocene system and these beds contain a quantity of remains of plants and animals. This formation stretches from the west border through Chichester and Arundel to Worthing and Brighton. The

Reading Beds are more prominent in the west, and the Woolwich Beds in the east. These rocks consist of red-mottled plastic or pottery clay, with seams of lignite, flint pebbles, and sand.

The London Clay lies over the last rocks : in Sussex it is more sandy than in the London Basin, and is of a dark blue colour. The Bracklesham Beds which succeed

View from the Devil's Dyke

the London Clay are among the most interesting deposits in the county. They are confined to the Selsey Peninsula, and have a thickness of about 500 feet. They consist of clays and marls, and contain remains of tropical animals and plants, such as turtles, crocodiles, sharks, palms, and pines.

The deposits above the Bracklesham Beds belong to

the most recent formations, consisting as they do of valley gravel, brick earth, blown sand, peat, and alluvium ; and with this brief notice of Nature's work which is going on in our own times we may fitly conclude our sketch of the geology of Sussex.

With reference to the soil of Sussex, it may be said that the Weald is generally fertile and richly wooded ; that the district westward from Brighton to the Hampshire boundary is a very productive tract, from two to seven miles broad ; and that in the south-east the rich marshlands afford excellent pasturage.

7. Natural History.

Various facts, which can only be shortly mentioned here, go to show that the British Isles have not existed as such, and separated from the Continent, for any great length of geological time. Around our coasts, for instance, are in several places remains of forests now sunk beneath the sea and only to be seen at extreme low water. Between England and the Continent the sea is very shallow, but a little west of Ireland we soon come to very deep soundings. Great Britain and Ireland were thus originally part of the Continent, and are examples of what geologists call continental islands.

But we also have no less certain proof that at some anterior period they were almost entirely submerged. The fauna and flora thus being destroyed, the land would have to be restocked with animals and plants from the

Continent when union again took place, the influx of course coming from the east and south. As however it was not long before separation occurred, not all the continental species could establish themselves. We should thus expect to find that the parts in the neighbourhood of the Continent were richest in species, and those furthest off poorest, and this proves to be the case both in plants

Sun Oak, St Leonard's Forest

and animals. While Britain has fewer species than France or Belgium, Ireland has still less than Britain.

Owing to the varieties of soil on the downs, in the Weald, in the forests, and along the seaboard, Sussex is one of the most interesting of our southern counties with regard to its flora. It is estimated that of 1960 species of plant-life in Great Britain, no less than 1159 are found in this county. The samphire, once so abundant, is still to be met with sparingly in all directions. Its collection once gave employment to cliffsmen, but is now a lost Sussex industry. The pretty rosy sea-heath (*Frankenia*) occurs along the shore in marshy flats, and the proliferous pink (*Dianthus prolifer*), one of the rarest flowers, is now found only near Selsey.

Among the most pleasing of the plants growing on the downs are the bell flowers—the pale blue harebell, and the clustered bell flower (*Campanula glomerata*) with its blossoms of a deep rich purple. Perhaps the most beautiful of all the plants growing on the chalk is the round-headed rampion (*Phyteuma orbiculare*) locally called the " Pride of Sussex." It occurs only in the south of England, and abounds on Beachy Head. Sussex is richer in orchids than any other county in England except Kent. A large number of the orchids love the chalk, many delight in the beechen " hangers," and others have their homes in the boggy lands at the foot of the Downs.

Sussex is a well wooded county, and further reference will be made to its great extent of forest land. The forests of St Leonard, Tilgate, and Worth, and the woods of Charlton, Goodwood, and Chilgrove have an abundance of trees. The fir, beech, birch, and pine are the principal trees, but in the Wealden district the oak is numerous. It is worthy of note that the yew is a specially common feature of the Sussex churchyards.

The wild animals of Sussex are similar to those in most English counties. The fox, stoat, and weasel are generally distributed throughout the county, the squirrel is abundant, and the otter is found in most of the large streams of the east. The badger fully holds its own in many parts, and near Hastings there is hardly a parish where he is not to be found.

Deer are very numerous in the many parks of Sussex, and probably no equal area of England contains so great

Arundel Castle

a number. The red deer are found in Buckhurst, Arundel, and Eridge Parks; while the fallow deer are still more numerous in the parks just mentioned, and in several others. The fallow deer of Sussex are second to none in size and superiority of their venison, while those in Petworth Park are said to be the finest in Great Britain. The roe-deer, the smallest and most beautiful of our deer, is, in Sussex, now found only in Petworth Park.

When we consider the birds of Sussex, we find that this county is more favourably situated to receive wanderers from the south, and spring migrants, than any other in England, as it is the first landing-place for all that come to spend the summer in our country. Hence many rare warblers and other birds have been recorded from the Downs in the Brighton neighbourhood, and its varied coast-line is naturally a favourite winter resort and resting-place at all seasons of a large number of species of water-birds. Among the smaller migrants are the wheatear, of which large numbers are found on the Downs in August and September. They are much esteemed as a delicacy, and are trapped in large numbers by the shepherds. Immense flocks of migratory wood-pigeons also visit Sussex.

Before closing this brief notice of Sussex birds, it is worth recording that there is a heronry in Parham Park. The number of nests has gone on increasing of late years, there being now about sixty. This heronry has a history. The ancestral birds were brought by Lord Leicester's steward, in the reign of Elizabeth, from Coity Castle in Wales, to Penshurst. There they stayed for 200 years, and then migrated to Michel Grove, near Arundel. About 1845, some of the trees in which they built were felled, and then the birds again migrated, and in three seasons all had found their way to the Parham woods.

8. Climate and Rainfall.

The climate of a district depends, among various factors, on the temperature, the prevailing winds, the dryness or moisture of the air, and the character of the soil; and the climate of a district may be defined as its state with regard to weather throughout the year. In considering the climate of Sussex, we must remember that it is a maritime county, having the modifying influence of the sea along its coast-line of upwards of 70 miles. It is also one of the largest counties, and consequently subject to more varieties of climate than we find in Surrey or in Middlesex. Again Sussex is further south than either of those counties, and so its latitude, to a small extent, leads us to expect that its climate will be warmer. Probably, however, its situation as a maritime county, and its sheltered position from the north and east winds, are the chief factors that give Sussex a warmer climate than that of Surrey or Middlesex.

It is now well understood that the climate of a county has considerable influence on its productions, and it is consequently of great importance to have accurate information as to the prevailing winds, the temperature, and the rainfall of a district. These, then, are some of the topics we shall now consider.

Our knowledge of the weather is now much more definite than it once was, and every day there appears in our newspapers a great deal of information on the subject. The Meteorological Society in London collects particulars from all parts of the country relating to the temperature of the air, the hours of sunshine, the rainfall, and the direction of the winds. The Meteorological Office divides the British Isles into ten districts for the purpose of information with regard to weather conditions for the twenty-four hours ending at midnight on the day when the news is published. Thus for February 3, 1908, the following was the forecast for Sussex, which is placed in the South England, London, and Channel District: " Light north-westerly breezes, backing westerly or south-westerly : fine and frosty at first, milder later, with local showers of rain or sleet; morning fog inland." When rough weather is expected, warnings are issued by the same office, and details are given on a variety of topics. Besides this official information, most of the daily news-papers print maps and charts in order to convey the weather intelligence in a more graphic manner.

For the collection of particulars with regard to rainfall another agency is at work. There are in the British Isles about 4000 observers who collect exact particulars of the rainfall in their locality. These results are arranged in a yearly record, known as *British Rainfall*, in which are entered the number of inches of rain that fell at various stations. In Sussex alone there are over 130 persons who keep a rain-gauge and enter in a register the daily rainfall. Every year these facts are tabulated for that station, and then forwarded to the editor of *British Rainfall*.

We are now in a position to consider some special facts bearing on the climate of Sussex, which lies in the

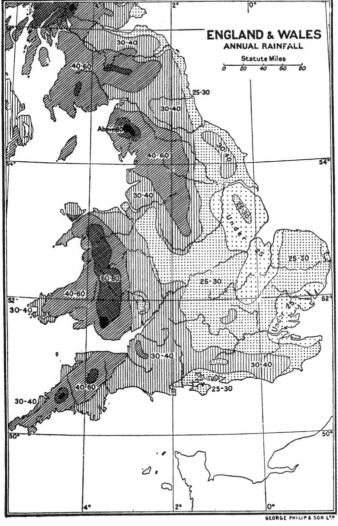

ENGLAND & WALES
ANNUAL RAINFALL

Statute Miles
0   20   40   60   80

30-40
40-60
25-30
30-40
Above 80
40-60
30-40
30-40
25-30
Under 25
25-30
60-80
40-60
30-40
Under 25
30-40
40-60
30-40
40-60
25-30
Under 25

GEORGE PHILIP & SON LT.º

(The figures give the approximate annual rainfall in inches)

3—2

sunniest district of the British Islands. Indeed, it has been calculated that it has about 1600 hours of bright sunshine in the year, out of a possible 4435 hours during which the sun is above the horizon. As we should expect, the sunniest places are on the coast, and Hastings, Eastbourne, and Brighton have each an average of upwards of 1800 hours of bright sunshine annually. These figures are higher than those of any other town in the south of England, with the exceptions of Weymouth, Falmouth, and Newquay. It is worth noting that, in 1899, one of the sunniest years on record, Bognor had no less than 2194 hours of bright sunshine. It is thus apparent that the Sussex coast-towns enjoy remarkably sunny skies ; and it is particularly noticeable in the winter that, when it is bright and sunny on the sea-coast, the country inland is cloudy and misty.

The prevailing winds of Sussex are from some point between west and south-west, and this fact teaches us that its climate is warmer than the north and east of England, although it is not so warm as the west and south-west. The mean annual temperature of Great Britain varies from about 46° in the north of Scotland to 52° in the Scilly Isles, while that of Sussex is about 51°. The difference between the mean temperature of the warmest month and that of the coldest month in Sussex is 23°. Of course the local climates vary considerably, but the South Downs form a dividing line between the two climatic districts of the county. The coastal district, south of the Downs, has a warmer winter and a cooler summer than the district to the north of that range. If we turn to the map just given we notice that, speaking generally, the rainfall decreases steadily as we pass from west to east. The moisture-laden clouds driven by the prevalent winds across the Atlantic precipitate their contents on reaching the land, more especially if the land be high, and in consequence the country beyond is less wet. Thus in 1906, the highest rainfall was at Glaslyn, in the Snowdon district, where 205·3 inches of rain fell; and the lowest was at Boyton Rectory, in Suffolk, with a record of 19·11 inches. The highest rainfall in Sussex, in 1906, was at West Dean Park, where 41·22 inches were registered; and the lowest was at Winchelsea with 24·01 inches of rain. It may be said generally that Sussex is wetter than the counties of the east coast, and not so wet as the counties of the west and south-west. The total rainfall for England and Wales in 1906 was 36·44 inches, while for Sussex it was about 32 inches.

Dr Mill, who has made a special study of rainfall statistics, says that "the whole of the low coastal plain up to the level of 100 feet has a rainfall under 30 inches.... The southern slope of the Downs and the whole of the valley north of the Downs have an average rainfall of from 30 to 35 inches; but the crest of the Downs and the narrow belt immediately at the base of the escarpment have a higher rainfall, closely approaching 40 inches." The rainfall of the watering-places differs slightly, and this is owing to the flatness or the hilliness of the coast. Thus while Eastbourne in the neighbourhood of Beachy Head had a rainfall in 1906 of 32 inches, Hastings and Brighton had each about 28 inches. Taken altogether, the driest months of 1906 in Sussex were April, July, and September, while January, October, and November were the months of greatest rain.

Eastbourne

9. The Coast—Gains and Losses. Its Protection—Sea Walls and Groynes. Lighthouses and Light=ships.

Sussex has a long seaboard stretching from Hampshire to Kent, and measuring at least 77 miles. A glance at a map will show that the sea coast is remarkably regular, and without a single harbour of importance. In this long extent of coast two points stand out and break the monotony. Selsey Bill, marking the termination of Selsey

Peninsula, is in the west, and Beachy Head, which is the extremity of some high land, is towards the east. Between these two points it will be seen that the sea coast forms a long and shallow curve, while beyond Beachy Head the coast trends generally to the north-east.

As we shall find in a later chapter, the coast of Sussex once had several important harbours and ports, and it has perhaps been more altered in form, and more filled up in the openings, than any other in England. This change has been brought about by the constant action of the winds and tides upon the materials within their reach, and one cannot but observe masses of shingle and sand lining the Sussex shore, which have been heaped up by these agencies upon strata of a kind different from themselves. Hence we may conclude that these masses of shingle and sand, swept along by wind and tide, together with the deposits of rivers, are the agents that have closed the harbours, choked the ports, and changed the form of the coast.

Now let us look at the map again, and note in order some of the features of the coast of our county. Leaving Hayling Island in Hampshire behind us we find that there is a shingle beach as far as Chichester Harbour. On the east of the harbour is a spit of shingle, and behind it a large bank known as East Pole Sand. Chichester Harbour cannot be entered at all at low water, and at no time without a pilot. Passing along to Selsey Bill, the cliffs of sand are as low as 12 feet or less, and are subject to erosion, wasting as much as from 6 to 8 feet in a year. In parts the land is below sea-level and is protected from the encroachments of the sea. The village of Selsey is half-a-mile from the sea, and was once the centre of a peninsula of which half has been washed away in 1000 years. Selsey Cathedral is beneath the sea ; and a deer-park which belonged to the Bishop of Chichester as late as the reign of Henry VIII is now an anchorage ground with three fathoms of water, and marked on charts as " The Park."

From Pagham (which once had a good harbour) to Bognor the beach is backed by a low earth-bank, and groynes are placed all along this coast. Indeed nothing is more noticeable on the Sussex beach in all parts than the numerous groynes, which are constructed to keep the beach level, and so to prevent the erosion of the coast. These groynes near Bognor are constructed of piles of fir, eight inches square, and spaced four feet apart, carrying thick planking to form a stout wooden wall, and running into the sea for a distance of from 50 to 100 yards. These are typical of groynes in other parts, although at Brighton and Hastings there are more elaborate groynes of stone and cement.

Between Bognor and Felpham a concrete sea-wall has been constructed for the protection of the low land. The coast continues low to Littlehampton, where the harbour is formed between two piers, which extend out across the beach for half a mile. From this place to Worthing there are groynes of various heights and sizes, placed at irregular intervals. When Shoreham is reached, we find that the drift of shingle has caused the Adur to be drawn out of its course. Shoreham Harbour was formerly of considerable importance, but although it is still busy in a small way, it has suffered from the changes which we have noticed in the early portion of this chapter.

From Shoreham to Hove the coast is low, and the beach is covered with shingle for some distance. The wasting of the cliffs at Brighton and Hove has been a

Worthing Sands

source of trouble and expense for three centuries, and now there is a most elaborate system of groyning. Indeed someone has remarked that "there are more groynes than beach." For the protection of the road and promenade in front of Hove and Brighton there is a line of sea-wall extending from Aldrington on the west to Black Rock on the east, a distance of four miles, and forming one of the finest promenades in England. The erosion of the cliffs has been very considerable, and the old road leading to Rottingdean became so dangerous that it was thought wise to stop it for passengers and open another further inland.

Brighton from the West Pier

From Rottingdean to Newhaven the chalk cliffs are from 80 to 100 feet in height, and at Newhaven and Seaford two chalk cliffs, on the west and east respectively, rise to 180 feet and 250 feet. Between these is the valley through which the Ouse finds its way to the sea. The coast now bends to the south-east, until Beachy Head is reached, the finest headland on the south coast. Beachy

Head is the termination of the South Downs, and is a precipice of chalk cliff over 500 feet high.

Passing Eastbourne, a fashionable and well laid-out watering-place, the coast gets lower, and a number of Martello Towers are noticed along the shore. These were built at the time when it was thought Napoleon would invade England, and it was considered that this portion of the coast was specially open to attack. Pevensey Harbour is now entirely filled with shingle, and the place is of no importance. St Leonards and Hastings are continuous towns, and both are protected by a sea-wall and promenade, three miles long, and by high groynes of timber or of concrete. The picturesque coast of Hastings gradually changes, and towards Rye it becomes low and flat. Both Rye and Winchelsea have "suffered a sea-change," and from being formerly important ports they are now mainly of historical interest. Rye, it is true, has some trade, both coasting and continental; but Winchelsea has fared badly. Old Winchelsea is beneath the waves, and New Winchelsea is left stranded inland one mile from the sea.

In bringing our survey of the Sussex coast to a close, we may pause to glance at the work that is done by the Elder Brethren of Trinity House to assist mariners in navigating our coasts by placing lighthouses, lightships, beacons, and buoys at various points. It is worth noting that a hundred years ago there were only about 30 lighthouses and lightships round the British coasts, and now there are about 900, of which 24 are in Sussex. The Elder Brethren of Trinity House derive an annual income of

£300,000 from dues levied on shipping, and this is used
for the purpose of lighting our coasts.

The earliest reference to lighthouses in Sussex is in
1664, when a licence was granted to improve Newhaven
Harbour, and to set up lights there and at Beachy Head.
The present lighthouse at Beachy Head is the finest in
Sussex, and stands at a height of 285 feet. The tower
itself is 47 feet high and its white light, which is shown
every 20 seconds, may be seen at a distance of 16 miles
in clear weather. Besides the lighthouse, fog-explosives
are used when required, storm signals are shown by the
Coastguard on Beachy Head, and there is telephonic communication between the lighthouse and the Coastguard.

Most of the sea-coast towns have lights on their piers
and jetties, and there is also one on the beach at Selsey
Bill. These are all lesser lights and are not comparable
with that at Beachy Head. There are two lightships off
the Sussex coast. The *Owers* light vessel on the west is
in 16 fathoms of water and has a red hull, with its name
on both sides. Its white and red light revolves every
minute and may be seen 11 miles away. On the *Owers*
there is a powerful fog reed-horn which gives one blast of
4 seconds every 10 seconds in foggy weather. The
other lightship is the *Royal Sovereign*, which is placed off
Pevensey Bay in 11½ fathoms of water at a distance of
three-quarters of a mile from Southern Head. It has a
red hull, with its name on both sides, and carries a small
ball placed over a large one at the mast head. Its white
light, which revolves every 45 seconds, may be seen at a
distance of 11 miles in clear weather.

10. People—Race, Dialect, Settlements, Population.

It is probable that the earliest inhabitants of Sussex were immigrants from the Continent when the British Isles were still part of Europe. We shall not be far wrong in assigning primitive man in Sussex to the period known as the Old Stone Age. It is generally agreed that Picts and others associated with the dolmens and other stone monuments succeeded the first inhabitants; and then came the tribes of Keltic speech, commonly called Kelts, who lived in the Bronze Age. There are few written records of these people till the invasion of the Romans in 55 B.C., when Julius Caesar found the Britons, or Kelts, belonging to various races, in different stages of civilisation, and using various modes of speech. The people of the part of England we now call Sussex were the *Regni*, a branch of the Kelts. The Downs and the Weald in the north, and the marshes about Chichester and Romney at the west and east, formed the boundaries of this tribe. There is every reason to believe that the Romans allowed the native chief to rule over his dominion, and so Sussex was left almost in its original independence.

The Romans built two strong forts in Sussex—one at Regnum, on the site of the present city of Chichester, and one at Anderida, where Pevensey now stands. The population of Sussex was to some extent augmented by the Romans and their legionaries. At the beginning of the fifth century, however, the Romans left Britain,

Market Cross and Cathedral, Chichester

and then Sussex fell an easy prey to the Teutons from the Continent. The Saxon conquest was so complete that, the English Chronicle tells us, there was not one Briton left. Be this as it may, we are certain that from that time Sussex became settled under one overlordship, and the Saxon settlers were very numerous.

The South Saxons were a people of Teutonic speech, and taken as a whole Sussex is one of the most Teutonic counties in England. From the fifth century onwards the English speech became general, and nearly all the places received new names, which have been retained to this day. The results of the Saxon Conquest were seen in the new language, in new laws, and in a conversion to heathendom.

After the settlement of the Saxons there were arrivals of Danes; and then in the tenth and eleventh centuries Normans, mainly Norsemen having a Romanised speech, came in considerable numbers. Since the Norman conquest, there have been frequent immigrations of foreigners from Europe. The Flemings in the time of Edward III, the French and Dutch protestants in the reign of Elizabeth and in the time of the Stuarts, settled in various parts of Sussex and intermingled with the native population. French influence was noticeable in Sussex in the Middle Ages, and such ports as Winchelsea and Rye had a constant influx of French people. The latter portion of the nineteenth century witnessed a steady arrival of foreigners from various European countries, and when the census of 1901 was taken, it was found that upwards of 6000 people in the county were of foreign origin.

From the foregoing remarks it will be gathered that the people of Sussex are mainly of Teutonic stock, and of English speech.   There are traces of Keltic, Norse, and French in the dialect.   For example the Hastings fishermen often say *boco* for plenty, and *frap*, to strike.   In the neighbourhood of Rye, where the Huguenots settled, such

Strand Gate, Winchelsea

words as *dishabil*, meaning untidy, undressed, and *peter grievous* (from *petit-grief*), meaning fretful, are still used. But, of course, the body of the Sussex dialect is of Saxon origin, and Saxon words meet us at every turn.   A cold wind is a *bleat wind*, a pig-stye is a *hog-pound*, and superior is *bettermost*.

We will now turn to some interesting facts relating

to the people of Sussex as we find them to-day. There
is no exact information with regard to the population of
our county till 1801, the year of the Union of Great
Britain and Ireland. Then the first census was taken,
and from that date onwards there has been a numbering
of the people every ten years.

When the first census of Sussex was taken in 1801,
the population was 159,471, and in 1901 it was 605,202.
This means that the increase has been nearly fourfold in
the century. During the last ten years the increase has
been upwards of 55,000, or about 10 per cent. on the
population of the previous decade. It thus appears that
the high rate of increase from 1871 to 1891 is not being
maintained. It is worth noting that considerably more
than half the increase during the century is due to
the growth of the watering-places, especially Brighton,
Hastings, and Eastbourne. The density of population to
a square mile in Sussex is 415, against 558 for the whole
of England and Wales.

The census returns of 1901 show that 399,182 people
live in urban districts, and 203,073 in rural districts; and
that the females exceed the males by 55,394. The Administrative County of East Sussex had a population of
450,702, or three times that of West Sussex. The bulk
of the people live in houses, or tenements, of which
94,649 contained five or more rooms, and 38,669 had
less than five rooms.

From the census returns we are able to gather particulars of the ages and occupations of the people. Thus in
1901, there were 37,850 people over 65 years of age; and

more than 7000 people were living in workhouses, asylums, and other public institutions. With regard to the occupations of the people, the men were chiefly engaged in agriculture, in house-building, as coachmen or servants, or as commercial men and clerks; while the women were mainly domestic servants, dressmakers, milliners, and teachers.

There is a very interesting table in the Sussex census that gives the place of birth of the people. Of the 605,202 persons, 389,147 were born within the county; 54,279 were born in London; 11,248 in Scotland, Ireland, and Wales; and 5414 in other parts of the British Empire. Persons of foreign birth numbered 6330, and were mainly natives of Germany, France, Italy, and Switzerland.

11. Agriculture. Main Cultivations, Woodlands, Stock.

Sussex is essentially an agricultural county and, as we shall find presently, more than two-thirds of the county are under crops and grass. In common with other agricultural counties, it has had its periods of depression, and since 1878 farmers have had uphill work to hold their position. In many parts of the county attention is now being directed to poultry-rearing and the cultivation of fruit on a large scale and according to scientific methods. This has been attended with the most satisfactory results, and to some extent has balanced the loss that has followed owing to the fall in the price of corn.

Arthur Young, a competent observer on agriculture, made a tour of some English counties at the close of the eighteenth century, and he has some very forcible remarks on the backwardness of cultivation in Sussex. He attributes this to the bad roads, and the small fields that were undrained and surrounded by woods and plantations. Since Young's day, however, a great change has taken place, for the roads have been improved, the land has been drained and limed, and good fences have been planted.

Probably the best time for farmers was from 1855 to 1877, when wheat, oats, beans, peas, and clover were the staple crops, and cattle and sheep were bred in large numbers. Wheat was sold at 50s. or more per quarter, whereas now it sells for 30s. or less.

Let us now consider the position of agriculture in Sussex at the present time, and to do this we will turn to the Report of the Board of Agriculture, which annually gives information as to the acreage and produce of crops, and the number of live stock in each county.

In 1905, there were 666,697 acres, or more than two-thirds of Sussex, under crops and grass. The "corn" crops were wheat, barley, oats, rye, beans, and peas, which were cultivated on 125,567 acres, or more than one-seventh of the entire area. Oats and wheat were the most important crops, the former accounting for 57,030 acres, and the latter for 50,440 acres.

The green crops comprise, among others, turnips and swedes, mangolds, cabbages, vetches or tares, and potatoes, and grow on 54,610 acres. Turnips, swedes and mangolds are the most important and occupy three-fifths of this acreage. About one-twentieth of the area of the county is devoted to the growth of clover, sainfoin, and grasses under rotation; and no less than 416,753 acres are under permanent pasture. This is by far the largest acreage, being four-ninths of the whole county.

The growing of hops has steadily declined in Sussex from 9989 acres in 1867 to 4647 acres in 1905. This decline is owing to a variety of causes, the chief being the superior character of Kentish hops, and the consequently lower price offered for Sussex hops, as well as the great increase in the imported article. The oast houses for drying the hops remain, but the land once devoted to this cultivation is now given over to other crops.

The cultivation of small fruit is much increasing, and at Worthing is a very important industry. In the neighbourhood of this town it is calculated that the greenhouses if placed end to end would stretch in a line for upwards of 40 miles. Grapes, cucumbers, tomatos, and strawberries are the chief crops, and fetch good prices in the markets to which they are sent, especially in London, Manchester, Liverpool, and Birmingham.

Much of the Sussex land once under cultivation is now laid down as pasture to produce milk for the large centres of population, such as Brighton, Hastings, and Eastbourne. The remainder of the area that is not under cultivation may be classed as bare fallow, mountain, and heath land, and forests, woods, coppices, and plantations, and these altogether account for at least one-sixth of the county. Sussex has from the earliest period been celebrated for its fine growth of timber, especially oak, which was long preferred by the naval authorities to that of any other district. In Saxon times the forest land of Sussex was part of the great Andredsweald which stretched from Hampshire into Kent. The most extensive forests are now St Leonard's, Ashdown, Waterdown, and Tilgate, and the chief trees are oak, ash, beech, Spanish chestnut, and birch.

Oxen at Work on a Sussex Farm

We will conclude our study of the Agricultural Report by considering the different classes of the domestic animals that are used for various purposes. The live stock of Sussex are classified as horses, cows and other cattle, sheep, and pigs, and of these sheep are the most numerous, accounting for 400,715 out of the total of 593,204 animals. Cows and cattle number 127,041, horses 24,346, and pigs 41,102.

The cattle are chiefly of the Sussex breed, and are unequalled for hardiness and beef production. Short-horns are bred largely for milk, and Jerseys for butter. The milk industry is most important, and there are large dairy farms and factories at Glynde and Sheffield Park. Pevensey Marsh is very fertile pasture land and is grazed by large numbers of cattle and sheep. The pasture land near Lewes, Newhaven, Rye, and Winchelsea, and by the Arun, also serves a similar purpose.

Oxen were once used for ploughing and were kept in large open yards. Teams of six bullocks used to draw the old wheel plough, and teams of eight oxen drew large wagons into the towns. Working oxen are now, how-ever, practically things of the past, and there are only about five or six farmers who use them for ploughing in the neighbourhood of the South Downs.

The Sussex sheep are among the best in the world, and the Southdown breed is unequalled for hardiness, good wool, and excellence of mutton. Although the number of sheep has somewhat decreased of late years, the price has improved, so that sheep-farming is very profitable.

Chicken-rearing is another profitable and improving agricultural industry, of which Heathfield is the centre. The fowls are known as "Surrey" fowls in London, where they fetch good prices, and in one week as many as 80 tons are sent to the metropolis. Bee-keeping is a cottage industry, and in the neighbourhood of the Downs an abundance of honey is produced.

12. Industries and Manufactures.

Sussex has no claim to rank either as an industrial or as a manufacturing county. It is essentially an agricultural county, and most of its industries are those connected, in one way or another, with agriculture. Most of our great industries are now carried on in the midlands, or the northern counties, and this is largely due to the fact that iron and coal are there found in great abundance. In Norman times, however, there was a very different condition of affairs, for neither coal nor iron formed an important item in English industry or trade, and the weaving trade was but little developed. Tin and lead were the chief mineral wealth, and raw wool and hides the principal articles of trade.

In this chapter we shall find that, on a smaller scale, a great change has also taken place in Sussex. Industries that were once important have ceased to be carried on, and other industries have succeeded them. The cloth industry, which was once widespread through the county, is now practically extinct. Broadcloth and kersey were made in many of the towns, and Chichester was an early seat of this trade. In the sixteenth century, weavers were to be found in almost every parish, and fullers and dyers are frequently mentioned. Not only was Chichester a centre of the cloth industry, but we find that, in the early eighteenth century, the spinning of linen employed many people there. Cambric goods were made at Winchelsea in the Middle Ages, and there is no doubt that this industry was introduced by the French, who settled at Winchelsea and Rye.

The timber industry has always been of considerable importance in Sussex, and this, of course, is owing to the extensive forests. Sussex oak has long been in demand, and in the Norman period it was used in the construction of the Tower and of Westminster Hall. It was also used at Portsmouth for the Royal Navy, and is now in

Shoreham and the River Adur

demand for plank-fencing and palings, and for the manufacture of wattles for sheep-farms. Before the seventeenth century, timber was exported from Shoreham and Rye in the form of billets for fuel, but in 1628 this was prohibited owing to the need of supplying the numerous iron-furnaces in the county. In 1901, there were 238 timber-merchants in Sussex, and 503 sawyers.

Ship-building was carried on at Hastings, Rye,

Winchelsea, Shoreham, and Arundel, and at the first sea-port some fine schooners were built for the Mediterranean trade. The ship-building trade has steadily declined, and in 1901 there were only 181 shipwrights and boat-builders at all the sea-ports. Brighton has the largest share of this industry, and is followed by Southwick, Eastbourne, Shoreham, Littlehampton, Rye, and Bosham.

Bosham

The manufacture of hoops for casks has long flourished in Sussex, and in 1901, no fewer than 284 persons were employed as hoop-makers and coopers. It is not generally known that the wooden baskets that gardeners carry are made in Sussex. Such is the case, however, and "trugs," as they are locally called, are associated with Hurstmonceux.

Among the other industries connected with Sussex owing to its abundance of timber may be mentioned tanning and charcoal. The tanning industry dates from the thirteenth century, and is now carried on at Chichester, Horsham, Battle, and Groombridge, on the borders of Kent and Sussex. Charcoal-burning was once of great importance, and large quantities of charcoal were exported. The demands of the iron-furnaces increased its value, but with the decay of the iron industry, its value declined. After a while, there was a revival in charcoal-burning, as charcoal was needed both in the making of gunpowder and also in the drying of hops. As early as the fifteenth century gunpowder was made at Rye, and there were powder-mills at Brede and Maresfield in the nineteenth century. In the year 1800, our Government set up a special establishment for the manufacture of charcoal at North Chapel, in order to supply the Government powder-mills at Waltham in Essex, and at Faversham in Kent. The North Chapel factory was closed, however, in 1831.

Paper-making was once more general in Sussex than it is to-day, where the only surviving mills are at Iping. The rope-making and sacking industry is now carried on at Hailsham, where twine, cordage, fibre mats, and hop-sacking are made and sent to all parts.

There are important cement works at Amberley, Newhaven, Upper Beeding, and Lewes, and plaster of Paris is made at Mountfield. There are breweries at Arundel, Horsham, Chichester, Brighton, and Lewes, and chemical works at Rye and Lancing.

In 1567, Jean Carré settled at Wisborough, and made

glass. This was a short-lived industry and came to an end in 1617. Bell-founding was once carried on at Hastings, Lewes, Slinfold, Lindfield, and Horsham, and many of the bells in the Sussex church-towers were made at one or another of these towns.

Among the industries connected with pottery, we may mention that roof tiles are made at Mayfield, Hastings, and Battle, and glazed pitchers and jugs at Horsham. The most important Sussex potteries are, however, at Chailey, where "rustic ware" of a peculiar shade of brown is made. This ware is remarkable for its ornamental work, which is mainly of green sprays and clusters of hops, acorns, leaves, and flowers, carefully modelled from nature. The Rye potteries turn out quantities of simple fancy articles.

We must not omit a brief notice of the bygone industry of salt-making. As the sea is the great source of salt, it is not surprising that this was once carried on all along the Sussex coast. The common method of obtaining salt was as follows. Sea-water was admitted into a number of broad, shallow "pans," or ponds with clay bottoms. The water was evaporated by the heat of the sun, and thus reduced to a strong brine. It was then boiled in shallow iron vessels, and allowed to cool, when the salt crystallised. In the Domesday Book of 1086, there is mention of 285 salt-pans in Sussex, and in the thirteenth century salt was sold to the French and Dutch in large quantities. There is now no need for this mode of obtaining salt, although there were extensive works in Sussex as late as the middle of the nineteenth century.

13. Minerals. Exhausted Mining Industries.

Sussex cannot now be considered a mining county although it was once a great iron-producing district. The ironstone of the Weald, both in Kent and Sussex, was very extensively worked until the establishment of the great iron and coal works in the midland and northern counties of England, which caused the industry to be abandoned. The ore used was the clay ironstone nodules found at the base of the Wadhurst Clay. These were dug in bell-pits of no great depth, and worked with oak charcoal. The result was a steely wrought iron of excellent quality. Further reference to the iron industry of Sussex will be made at the end of this chapter.

The quarries of Sussex are now little worked, but there is a most interesting variety of stone known as "Sussex marble." It is a calcareous stone, formed by a deposit of freshwater shells, and takes a high polish. It is frequently used for ornamental purposes, such as chimney-pieces, and for building, paving, and burning into lime. Much of it was employed in building Canterbury Cathedral, where it is called Petworth marble, being found in the neighbourhood of that town in the highest perfection. It was also used in Chichester Cathedral and the building of Petworth House.

The limestone and the ironstone in contact with the "Sussex marble" often rise within a very few feet of the surface. Alternate strata of ironstone and sandstone occur almost everywhere in the Weald; and under these, at a considerable depth, are strata of limestone, which, when burned, makes the best cement. The sandstone was worked by the Romans at Pulborough for use at Bignor; and a quarry of greensand at Eastbourne supplied the stone for the Roman station at Pevensey, and for building and repairing the castle at that place. The sandstone of Sussex was also largely used for building churches, and

Pulborough Church

for iron-furnaces. At Horsham and Slinfold there are some important quarries, and the stone, which can be easily split, is used for roofing purposes. Of late, however, it has fallen into disuse on account of its weight, and the expense of carriage.

Chalk is largely used in Sussex for building purposes,

and for conversion into lime, which is used either for the making of mortar, or as manure. In the eighteenth and nineteenth centuries nearly every large farm near the Downs had its own kiln, and lime-burning was an industry of some importance. Arthur Young at the close of the eighteenth century describes a kiln, at Hastings, having a capacity of 1200 bushels, and a daily yield of 300 bushels. In 1851 there were 66 persons engaged in quarrying and in lime-burning, but in 1901 this number had fallen to 47. This decrease is due to the fact that lime is not so freely used on the land as it was at one time. The principal lime-works are at Amberley, Glynde, Lewes, Pulborough, and Jevington.

Among the other mineral products of Sussex may be mentioned gypsum, which was struck at Mountfield in 1872. Beds of red ochre are found at Graffham and Chidham, and fuller's earth occurs at Tillington. Flints are collected from the chalk-pits and from beneath the turf on the Downs, and used for building purposes and for road-making. When first dug the flints are too brittle for use, and it is quite a common sight to notice large heaps of flints spread out to weather. After some exposure they become tough and durable. The brick-earth at Littlehampton, Rustington, Worthing, and elsewhere is very suitable for making bricks. Sussex bricks have a good name for their warm red colour, and the best are made at Ditchling, Keymer, and Burgess Hill. Here it may be mentioned that Hurstmonceux Castle is one of the earliest brick edifices in England, and the most picturesque ruin in Sussex.

Hurstmonceux Castle

At the beginning of this chapter we referred to a period when Sussex formed part of the Black Country of south-eastern England ; and it is interesting to note that in the sixteenth and seventeenth centuries the iron-works of the Weald were by far the most considerable in all England.

The industry flourished in the Weald because of the almost inexhaustible supply of timber which could be converted into charcoal for fuel in the iron-furnaces. In 1319, the counties of Surrey and Sussex were ordered to provide 3000 horseshoes and 29,000 nails for an expedition against the Scots. During the fifteenth and sixteenth centuries the iron-works increased in importance, largely owing to the use of cannon in war ; and in 1543 the making of cannon had become a notable part of the industry in Sussex. In that year the first cannons cast in one solid piece were made at Buxted, by Ralph Hogge ; and there may yet be seen some old Sussex banded cannon in the Tower.

By the middle of the sixteenth century it was found that the iron-works of the Weald were consuming so much timber that it was necessary to pass an Act forbidding timber to be cut down for iron smelting within 14 miles of the coast. This was the first check the iron industry had received, and there is no doubt that the Government viewed the industry with some suspicion, as it was feared that the supplies of oak from Sussex for ship-building would be stopped. The great woods in the Ashdown district have disappeared, and the South Downs are now bare and treeless, so that we can form some idea of the enormous quantities of wood that must have been consumed.

Notwithstanding many prohibitions, iron smelting continued to flourish until the Civil War, when the iron-works belonging to the Crown and the Royalists were destroyed by Waller after the sieges of Chichester and Arundel in 1643. However, the industry lingered on for more than a century and a half, and the last furnace was at Ashburnham in 1828.

The chief centres of the industry in Sussex were at Lamberhurst, Maresfield, Buxted, Mayfield, Ashburnham, and Penhurst. Lamberhurst boasted that it had the largest furnaces, and made the biggest guns, and it was at this place that the massive iron railings that surrounded St Paul's in London till 1874 were made. These rails were 2500 in number, and, with seven massive gates, weighed 200 tons and cost over £11,000. The oldest existing iron article made in Sussex is a cast-iron monumental slab in Burwash Church, made in the fourteenth century. There are many other iron monuments in various Sussex churches and churchyards, and andirons and chimney backs are to be seen in old mansions and farm-houses. Traces of the iron industry may be found in the names of ponds and places, such as Furnace Pond, Forge Pond, Hammer Pond, Horseshoe Farm, and Cinder Hill. Hammer Ponds are numerous both in Surrey and Sussex, and were formed by damming up the streams and using the water to turn water-wheels, which lifted up and let fall heavy hammers.

Hammerpond Waterfall

14.   Fisheries and Fishing Stations.

Our English fisheries, especially those on the south and east coasts, are of considerable importance, and employ many thousands of people. The amount of capital in the fishing industry is very large, for vessels and boats are costly, and expensive machinery is carried on board. The vessels and gear are subjected to very hard wear and sometimes both vessels and gear are lost altogether. Since the introduction of steam, the capture of fish and its consumption have greatly increased. Before the age of steam very little of the fish found its way beyond the coast towns, where it was sold by the fisher-folk from house to house. Now the fish is no sooner landed than it is packed, and carried by the railways to all parts of our country.

As we might naturally expect from its extensive coast-line and numerous small harbours, Sussex has always been an important centre of the fishing industry. Even during the period of the Roman occupation the shell-fish of Sussex were common articles of food ; and at the Roman settlement of Pevensey, or Anderida as it was then called, extensive waste-heaps of oyster, cockle, and mussel shells have been discovered.

It is also of interest to note that the conversion of Sussex to Christianity was largely owing, according to Bede, to the improvement of the fishing industry by Wilfrid, "the Apostle of the South Saxons." Bede's story is so interesting that it is worth a passing notice.

5—2

It was in 681 when Wilfrid landed in Sussex, and found the people so truly barbaric that they were ignorant of fishing except for eels, although the sea and rivers abounded with fish. Wilfrid, however, bade his attendants collect nets used in eel-fishing, and cast them into the sea. Presently they hauled in three hundred fish of different sorts, which they divided into three parts,—for the poor, for the lenders of the nets, and for themselves. Bede then tells us that by this "good service, the prelate turned their hearts powerfully to love him, and they were the readier to listen hopefully to his preaching about heavenly benefits, after they had through his agency received temporal good." It is also worth recording that Wilfrid became the bishop of the South Saxons and established his seat at Selsey.

In the thirteenth century, Winchelsea was well known as a fishing port and supplied the King's table with fish, while in the next century it is recorded that Winchelsea plaice and Rye whiting were held in high esteem. The Winchelsea fishermen not only pursued the industry on the south coast, but they also manned 14 ships to take part in the Yarmouth herring fishery. Early in the seventeenth century, the intrusion of French ships into the Sussex fishing grounds gave much trouble, and in 1622 a number of French vessels were captured off Rye. In the eighteenth century, the Sussex fisheries were fairly flourishing, and we find that considerable smuggling was combined with the fishing trade. It was at this period that the Brighton mackerel fishery brought in large sums of money.

In our own time, with changed conditions, it was recognised that the sea fisheries of England wanted regulating, and in 1888 an Act was passed which constituted a Sea Fishery District for Sussex, extending from a line drawn south-east of Dungeness Light to a line off Hayling Island. By this Act a Board of nineteen members were appointed to supervise this fishery district.

The Dieppe Boat leaving Newhaven

The fishing ports of Sussex are now Rye, Hastings, Eastbourne, Newhaven, Brighton, Shoreham, Worthing, Bognor, and Selsey. Winchelsea has no boats, but conducts its fishery with kettle- or stake-nets. These consist of fences formed of hurdles, nets, or stakes, which are completely covered at high tide, so that fish can swim over and round the fence; but with the fall of the tide, the fish are cut off and captured. This method is of great antiquity.

The other methods are various and are adapted to the different kinds of fish. Thus at Rye and Hastings, trawling and drift-nets are used for catching soles, plaice, herring and mackerel, while lines are used for cod, turbot,

Beachy Head

whiting, etc. The Eastbourne fishermen have drift-nets and long lines for mackerel, plaice, and skates, and employ pots for crabs, lobsters, and whelks. Selsey fishermen use drift-nets for herrings, lines for skate and cod, pots for whelks, lobster, crabs, and prawns, hand-fishing for periwinkles, and dredging for oysters.

The fishing industry of Sussex is worth about £60,000

a year. This income is derived chiefly from the capture of shell-fish, especially in West Sussex, where the lobsters, prawns, and cockles of Selsey have long been famous. Oysters are dredged off Selsey, and are cultivated at Bosham and Emsworth. Until 1870, there were oyster-beds at Eastbourne and Pevensey, but owing to various causes they no longer exist. Arundel mullet are of well-earned celebrity, and the best grey mullet in England are caught in the Arun.

Freshwater fish were once of far greater importance, and in the Domesday Book there are references to fisheries in Sussex. In 1798, Arthur Young wrote that carp were the chief stock, but that tench and perch, eels and pike were raised. Nowadays sea-trout, grilse, and sometimes salmon are caught in the lower Ouse, but the catching of freshwater fish is no longer an industry, but a sport.

15. Shipping and Trade—The Chief Ports. Extinct Ports. Cinque Ports.

Although Sussex has a long coast-line, it has very few sea-ports and none of first-class importance. As we have seen, few if any parts of the English coast have suffered so many changes, for several Sussex ports that once had good harbours and carried on an extensive trade have either been closed by the eastward drift of shingle, or have been forsaken by the sea.

There are now only three towns that have any pretension to be called sea-ports. Newhaven, at the mouth of the Ouse, is the most important, and is the only harbour of any note between Portsmouth and the Downs. It has a line of mail-steamers that run to Dieppe, and there is considerable trade with North France, especially with Honfleur and Caen, and also with the Channel Islands. Newhaven has also a coasting-trade in corn, coal, and timber. Rye, on the Rother, has a very limited trade, which strongly contrasts with its flourishing condition in the Middle Ages. It now exports wool, corn, timber, oak-bark, and hops, and imports coal and manufactured goods. Shoreham, at the mouth of the Adur, is now almost extinct as a sea-port, for most of its trade has been transferred to Newhaven. In the bright days of its prosperity great cargoes of corn and wine were landed here, and in the time of Edward III it provided 26 ships for the invasion of France.

The extinct ports of Sussex are quite numerous, and this county shares with Kent the barren honour of having the largest number of decayed sea-ports in England. We have already alluded to the reason for this decay, so that we may now devote some attention to the past history of these places. Perhaps we shall do well to consider them with the Cinque Ports, which were situated in Kent and Sussex. As the word implies, the Cinque Ports were originally five in number, but afterwards two others were added to the confederation. The original ports were Hastings, Sandwich, Dover, Romney, and Hythe, and the two "Ancient Towns," as they were called, were Rye and Winchelsea. Of these Seven Head Ports, it will be seen that Hastings (the premier Cinque

Port), Rye, and Winchelsea were in Sussex. Besides the Seven Head Ports, there were 32 other ports, termed 'limbs,' attached to them, and of these, the following were in Sussex:—Seaford, Pevensey, Bulverhythe, Hydney, and Ilam. So completely has the sea done its work, that in some cases the very site of the ports is now not known with certainty.

Ypres Tower, Rye

The Cinque Ports were originally a corporation to control the fisheries, and to provide for the defence of the south-eastern coasts. They thus were a sort of local Royal Navy, and as late as the reign of Henry VII they provided many ships, while in Elizabeth's reign they came to the front and helped to defeat the Armada. It is said that, at that great crisis, every man in the ports sprang to his post and watched the coast.

In return for providing ships and men to serve the King for a certain period in each year, the Cinque Ports had many privileges. They were allowed a certain form of self-government, their freemen were permitted to trade free of toll in all English boroughs, and the men were exempt from military duty. In addition to these privileges the "Barons" were honoured with the highest place at Coronations, and the Warden had command of the navy that guarded the southern shores of England. In the fifteenth century the Cinque Ports began to decline, and as King Henry VII formed a new Royal Navy, their assistance was no longer needed.

Hastings, the Premier Port, as we know it to-day is the third town of that name and was built about the middle of the seventeenth century. The site of the first town is under the sea, and there is now not even a trace of an inlet on the coast-line. Rye has not fared so badly as some of the other Cinque Ports. In the fifteenth century it combined with Yarmouth in the fishing trade in the North Sea, and its fleet engaged in the wine, timber, and billet trade. Its former importance is shown by a reference to its charter, which concludes with the words, "God save Englonde and the Towne of Rye"; and when Queen Elizabeth visited it in 1573 she named it "Rye Royal." Winchelsea has suffered more than either of the other Sussex Head Ports. The old town lies beneath the sea, and the new Winchelsea, built by Edward I, has been left a mile inland. When

Elizabeth visited it in 1573 she named it "Little London," perhaps in jest, for in 1601 we are told that Winchelsea had " gone to decay."

Hastings Castle : the Chancel Arch of the Chapel

Seaford was a " member " of Hastings and showed signs of decline in the reign of Edward III.   The course of the Ouse, blocked by shingle from the west, was diverted, and in the sixteenth century a harbour, " New Haven," was formed at its new mouth.　Pevensey, once a convenient port, and even the rival of Hastings, is now only a hamlet, with the ruins of its castle to tell of its former greatness.　Bulverhythe was near Hastings, Hydney close to Eastbourne, and Petit Ilam in the vicinity of Winchelsea.　These and other sea-ports have long since disappeared, and are only remembered in history or by local traditions.　Thus when the sea makes a kind of rattling sound on the shingle, the fishermen say they "hear Bulverhythe bells," and expect rough weather.

16.　History of Sussex.

When Britain was invaded by the Romans, Sussex formed part of the territory of the Regni, a British tribe. The Emperor Claudius commissioned Flavius Vespasian to subdue this part of the island, and about the year 47 A.D. the Roman dominion was established in the maritime portions of Sussex.　This territory was included in the division called Britannia Prima, and the two seats of Roman rule were at Regnum, probably our Chichester, and at Anderida, afterwards Pevensey.　It appears that the Roman rule in Sussex was mild, and that the native chief was allowed to retain his power over his dominions. During the fourth century the Saxon pirates made many attacks upon our southern and eastern shores, and in order to protect these parts the Romans placed strong garrisons at seven fortified places between the Wash and Beachy Head. All these seven towns were placed under an important Roman commandant, or Count of the Saxon Shore (*Comes Littoris Saxonici*) as he was called. Anderida, seven or eight miles from Beachy Head, was the fortified sea-port in Sussex, and the remains of its Roman wall yet testify to its former importance.

There is no particular mention of Sussex until after the departure of the Romans. In 477 A.D. four Saxon chieftains landed in Sussex, and the *English Chronicle* is the only authority for our information. It says, "Aelle with three sons and three keels come to the place called Cymenes once. He slays many Britons and drives others to take refuge in the wood that is called Andredsweald." For seven years after their coming, the Saxons kept to the western half of the county, probably in the neighbourhood of their new capital, Chichester, which had been renamed after one of Aelle's sons, Cissa. In the eighth year, they again fought with the Britons, and in 491 A.D. advanced as far as Anderida. Let us turn again to the *English Chronicle*, which tells the story in the following sentence : "Aelle and Cissa attacked Andredesceaster and slew all who dwell therein nor was there for that reason one Briton left alive." From that time, it is probable, the whole of Sussex became united under a single ruler, whose chief seat was at Chichester.

The first king of Sussex was this Aelle (Olla, or Ella), a Bretwalda, and in 514 Cissa succeeded his father. During the next hundred years, Wessex was taking rank as the chief English kingdom, and in 607 Ceolwulf of

Wessex conquered and absorbed Sussex into his dominions. Although the South Saxons were subject to Wessex, they retained their own kings for many years. In 823 Egbert, King of Wessex, gained the submission of the kingdoms of Kent, Surrey, Sussex, and Essex, and when he died in 836 he left Sussex as the portion for his son Athelstan. Later we find that King Alfred, who had large possessions in Sussex, was living at West Dean, near Seaford, and here he met Asser, who became the biographer of the great king. In Alfred's reign the Danes attacked Sussex, landing near Chichester. Many of them were killed, and their ships were captured. In 893 they landed in the eastern part of the county, and proceeding up the river Rother seized the town of Appledore, in Kent. The famous Danish pirate, Hasting, also landed near the site of the present town of Hastings, where he raised some fortifications.

The tenth century is almost a blank as far as Sussex history is concerned; but in 994 Olaf of Norway and Sweyn of Denmark, having failed in an attempt to capture London, turned into Sussex and "wrought the most ill that ever any army could do in burning and harrying, and in man-slaying." Peace and rest were given to Sussex by the Danish King Cnut; and after awhile Godwine had one-third of Sussex and lived at Bosham. It was from this little port that Harold sailed on his excursion to Normandy, when he fell into William's power.

The year 1064 is supposed to be the date of the visit of Harold to the court of William the Norman. The English Earl was forced to make an oath of fealty to the

The Gateway, Battle Abbey

Norman prince, and it was the breach of this oath on the death of Edward the Confessor that brought about the invasion of England by William. Harold was crowned King of England in the early months of 1066, and at once the Norman preparations began, and lasted on through the summer and almost up to the middle of autumn. Harold also collected his fleet and army, and in the words of the old Chronicler, "it was such a force both by land and sea as no king of the land had ever gathered before." Harold went to the Isle of Wight and there lay at anchor all the summer, but as his provisions were exhausted, and possibly thinking all danger of invasion was over, he allowed his men to go home.

In taking this step Harold made a great mistake, for he was soon threatened by an unexpected invasion in the north by Haardrada, King of Norway. It is not necessary to go into the details of the fights at Fulford and Stamford Bridge, but in passing it may be noted that no sooner had Harold defeated his northern foes on September 25, 1066, than he heard that William of Normandy had landed at Pevensey. It is related that the Norman ships numbered 696, and that their landing on the morning of September 28, was unopposed. The tidings of William's landing was swiftly carried northwards, and reached Harold at York on October 1. The English king at once set out for London and by October 12 had collected his army and marched southward. He took up a position on the last spur of a low range of Sussex hills about seven miles to the north-west of Hastings. It was on the morning of Saturday, October 14, that Harold's army

was drawn up in line on the ridge now crowned by the
abbey and town of Battle, and William's army faced
them on the hill which now bears the name of Telham.
The battlefield is sometimes called Senlac, but it will
probably always be best known by the name of Hastings.

Harold's position on a hill 260 feet above sea-level,
and surrounded by narrow valleys, was very difficult of
approach by cavalry. It was made yet more secure against
such an attack by a fence or palisade, as well as by a fosse
drawn right across the field. William's troops probably
numbered about 15,000, and Harold's were rather fewer.
The Norman troops were better trained and seasoned than
the English, but Harold had undoubtedly the advantage,
for his army stood on the defensive in a position which
had been chosen with considerable skill.

The two armies stood fronting each other in battle
array at 9 o'clock on the morning of October 14. The
men of Kent claimed the right to be in the van of the
English army, and strike the first blow in the battle, and
the Londoners made a claim to guard the king, being
grouped round his standard which was placed in the
middle of the ridge. William put his archers in the
first line of his army, his mail-clad infantry in the second
line, and behind them all he placed his cavalry. The
Normans were in the centre, the Bretons on the left, and
the Frenchmen on the right.

For six hours the battle raged with nearly equal
fortune on both sides, but afterwards victory inclined to
the Normans, chiefly owing to William's strategy in
feigning flight and drawing the English from their strong

position on the hill. William also bade his archers to shoot high up into the air, and one of this terrible flight of arrows struck Harold in the right eye. He fell to the ground mortally wounded and was soon slain by four Norman knights. Although the battle was continued for some time, all was in favour of the Normans, and ere

Battle Abbey: Site of the High Altar erected on the spot where Harold fell

the evening came, William's banner was planted on the brow of the hill where Harold's had lately floated.

William, in fulfilment of a vow made on the eve of the fight, founded on the field of battle a stately abbey which was named Battle. The building of the abbey must have done much to alter the face of the battlefield;

and now for nearly four centuries the abbey has been hidden and changed by the manor-house built in Tudor style after the suppression of the monasteries. It is, however, still possible to point to the site of the high altar of the Abbey Church on the crest of the hill as the spot where King Harold fell.

The Battle of Hastings was one of the decisive battles of the world, and was a turning point in our national history. The story of England as ruled by Anglo-Saxon kings ends, and a new chapter opens when the great Norman Duke becomes also the King of England.

After the Battle of Hastings, the history of Sussex presents no striking features. William II besieged Pevensey Castle in 1087 and after a severe siege of six weeks compelled the garrison, at whose head was Odo, Bishop of Bayeux, to surrender. In the troublous reign of Stephen the two castles of Arundel and Pevensey were held for Matilda against the king. The worst of our kings, John, was often in Sussex, either going to, or returning from France, but the only Sussex event of note in his reign was the rebellion of William de Braose of Bramber Castle.

The decisive battle of Lewes, fought on May 15, 1264, was between the forces of Henry III and those of his barons under Simon de Montfort. Both the king and his son, Prince Edward, were made prisoners, and for a time Simon de Montfort had considerable power. In the peasants' rising of 1381, Sussex played a prominent part, and Jack Cade, leader of a rebellion in 1450, was killed by Iden, Sheriff of Kent, in a garden near Heathfield.

Sussex took a worthy part in preparing for the descent

of the Armada, and raised 4000 foot soldiers, besides 260 horse soldiers for the defence of the coast. When the Armada had passed the Sussex coast on July 28, 1588, the county forces were dismissed to their homes.

The Civil War did not affect Sussex to any great extent. East Sussex seems to have sided with the Parliamentary forces, while the Royalists were strong in West Sussex. In 1642, Waller besieged and took Chichester, and later both Arundel and Bramber Castles were taken. Soon after this Prince Charles was defeated at Worcester, and at the end of his wanderings reached Brighton on October 14, 1651. Next day he escaped by sea from Shoreham and set sail for Fécamp.

On June 30, 1690, the combined fleets of the English and Dutch were defeated off Beachy Head by the French; and at the beginning of the nineteenth century the coast was fortified owing to an alarm of invasion by Napoleon. During 1803–4 Martello towers were built, and the military canal from Rye across Romney marsh was dug. Fortunately these precautions were unnecessary, and to-day the towers or their ruins only remind us of the fear of the Napoleonic invasion more than one hundred years ago.

17. Antiquities — Prehistoric. Roman. Saxon.

The earliest written records of Sussex and its people do not carry us back more than 2000 years ago, so that prior to that period, and even for some time later, we are dependent for our knowledge of the dwellers in Sussex on the traces they themselves have left of their handiwork.

Antiquaries have divided the earliest period of our country's history into the Stone Age, the Bronze Age, and the Iron Age. The Stone Age has been subdivided into two periods, the first being named the Palaeolithic, or Old Stone Age, and the other the Neolithic, or New Stone Age. The Stone Age is distinguished as the period when early man shaped flint implements by flaking, chipping, crushing, and grinding. We cannot here state the process by which flint in its natural state was chipped and broken into such implements as hatchets, knives, hammer stones, and other forms that were needed by primitive man in Britain, but the greatest deftness and ingenuity were displayed in making these stone implements or weapons.

It is generally easy to distinguish the earlier, or palaeolithic, implements from those of late date by the bold style of workmanship. The various implements of this class were probably the very earliest of man's attempts at tool-making, and the character of the work shows that the workers were men of some skill and intelligence. Implements of the early Stone Age have been found in many parts of Sussex, especially at Friston near Eastbourne, East Dean, Brighton, and Midhurst.

The Neolithic, or Late Stone Age, is well represented in Sussex, both with regard to stone implements and earthworks. The work of this Age is of a more advanced type, and there is evidence of remarkable power of producing straight, uniform, and nearly parallel fractures in the flint. The workers of the Neolithic Age selected flint of

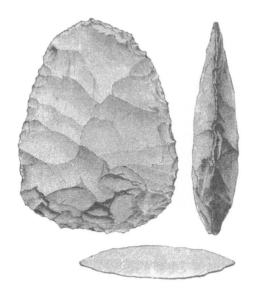

Palaeolithic Implement
(From Kents' Cavern)

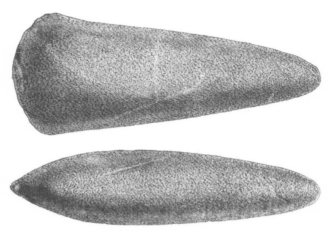

Neolithic Celt of Greenstone
(From Bridlington, Yorks.)

a superior quality, and developed the art of delivering blows on the flint with a fine precision. In this later Stone Age the following implements and weapons were made:—picks, chisels, flakes, scrapers, borers, arrow-heads, hammer stones, and grinding stones ; and of these scrapers, simple flakes, and cores have been most found in Sussex, especially on the south slope of the Downs. Celts and hatchets, highly finished, have been unearthed at Cissbury Hill, and various flint implements at Eastbourne and Pevensey.

Numerous earthworks and hill-top camps on the highest points of the South Downs are believed to belong to the Neolithic Age. Perhaps the best example is Cissbury Ring, 603 feet high, having on its summit an oval entrenchment, the largest and finest on the South Downs. Here was a flint manufactory, and the highly finished implements discovered in the neighbourhood show that the people had considerable skill.

The Bronze Age marks a period when the people of our land began to learn how to fashion metal, make pottery, and form defensive earthworks. It must not be supposed that only bronze was used in this period, for we are sure that gold was known, and even worked. Perhaps for our purpose we shall do well to group together the Bronze Age and the Iron Age, and call them the Age of Metals. The antiquities of this period have been found as separate objects, as hoards, and as sepulchral deposits. When bronze was introduced into this country is not certainly known, but there is reason to believe that the Kelts made the earliest specimens of bronze implements and weapons.

There are many barrows of the Bronze Age in Sussex, and those near Eastbourne, at Lewes, Brighton, and Chichester have yielded many remains of this period. At Hollingsbury Hill, near Brighton, a remarkable series of antiquities has been unearthed, and remains of pottery have been found at Alfriston, Hove, and Lewes. In the interior of a barrow near Hove there was found a rude oaken coffin containing, among other objects, a bronze dagger and an amber cup. In 1863, there was a noteworthy discovery of gold ornaments at Mountfield, near Battle. A ploughman turned up a long piece of metal and a great number of rings, weighing 11 lbs. avoirdupois. He sold these for old metal and received 5s. 6d. It was afterwards discovered that they were gold ornaments of the Bronze Age, and they realised in London no less than £529.

There is a curious figure known as the Long Man of Wilmington that probably belongs to this early age. It is an outlined figure 230 feet long cut in the turf of the South Downs, which may be seen from the railway near Polegate. This gigantic figure, which is cut on the slope of the hill, holds an immense staff in each hand. There are other similar figures, but mostly in the shape of horses, to be seen in various parts of England. Most of them are of this age.

Hoards of British coins of gold, silver, copper, and lead have been found at Battle, Bognor, Lancing Downs, Selsey, and Ashdown Forest. Artificial caves in the chalk may be seen at Hayes Down and Lavant, and in them have been found neolithic implements, and pottery of the Roman period. St Clement's Caves at Hastings are

extensive and of uncertain origin, but may belong to
this period. Ancient boats of the British period have
been found at Bexhill, Burpham, and North Stoke.

The Roman relics in Sussex are interesting and important, though not so numerous as those in the neighbouring county of Kent. Chichester, the ancient Regnum,
the capital of the Regni, has many traces of Roman

The Long Man, Wilmington

occupation, the most important of which, the "Neptune
and Minerva Slab," is preserved in the Town Hall. The
Roman walls of Anderida still remain at Pevensey, and
there are no less than nine towers at intervals in the wall.
The chief Roman treasure of Sussex is the villa at Bignor
with its large and remarkable pavements. The buildings,

East Lavant Church

discovered in 1811, cover a space of 650 feet by 350 feet, of which the villa itself with its court occupies one half. The Roman road, Stane Street, runs close to the villa, and owing to this fact it is probable that some wealthy Roman official chose this site for his house.

As might be expected, the Saxon remains in Sussex are very numerous. At Mill Field, near Eastbourne, many graves a few feet below the surface were opened, and knives, spears, shield bosses, glass cups, wooden buckets, armlets, and swords of the Saxon period were discovered. In a Saxon cemetery near Lewes some skeletons with weapons of iron (swords and spearheads) were found; and graves of women were unearthed in which were beads of glass, amber, and amethyst, brooches, armlets, buckles, and coins. Saxon coins have been discovered at Alfriston; and at Chanctonbury, in 1866, thousands of coins were found, which had been struck at Chichester, Hastings, Lewes, and Steyning.

18. Architecture. (a) Ecclesiastical— Churches, Cathedral, Abbeys.

Sussex has not so many fine churches as the neighbouring county of Kent, or as Norfolk and Suffolk, but it is far richer in these and other ecclesiastical buildings than is generally supposed. The churches are of all periods from Pre-Norman times, and of the most diverse character, but the greater number of them deserve careful attention.

Before considering them, however, we must first recall

Malling Hill, Lewes

the fact that Christianity was established in Britain, and of course in Sussex, during the Roman occupation. Here and there, we still find traces of Roman influence in some of our oldest churches. After the Romans left our country, the Saxon conquest brought a revival of heathenism, and so pronounced was the falling away from Christianity in Sussex that it was not till 681 that Wilfrid was the means of converting the South Saxons to the true faith. In that year he was received by King Ethelwold, who, with his officers and chief men, was baptised. The king gave Wilfrid 87 hides of land in Selsey, on which were 250 slaves, all of whom were at once set free. Wilfrid became the first bishop, and fixed the see at Selsey, where it remained till 1075, when it was removed to Chichester, of which diocese Stigand was the first bishop.

For more than 1000 years we can trace the progress of ecclesiastical architecture in Sussex, and there are at the present time at least 313 old parish churches which date from Pre-Reformation days. Besides these buildings, and Chichester Cathedral, there are remains of some of the religious houses, so that the subject of ecclesiastical architecture in Sussex is extensive.

The character of the buildings depends largely on the materials accessible. Speaking generally, the stone found in Sussex is not of the best quality for durability or for good masonry. The stone used in some of the churches, especially those near the sea-coast, was imported from Caen, in France, and it is a curious fact that this building stone was sent in exchange for cargoes of Sussex wheat. Some of the churches along the sea-coast,

The Porch, Bishopstone Church

and in the western and central districts, are largely built of flints obtained from the sea-shore or from the chalk. In East Sussex sandstone was used in many churches, and in the north-west part of the county chalk-rag is common in some buildings. The chalk from the Downs was used for rubble and interior work; and the hard chalk-rag for the walls. Clunch was freely used, for it was easily worked, and hardened to exposure.

The sandstone of Pulborough was of brown, yellow, greyish-green, and orange tones, and was used in the churches at Pulborough, Arundel and Lyminster. Horsham stone was employed for roofing, and Sussex, or Petworth, marble may be seen worked into the interior of Chichester Cathedral. On the borders of Surrey much timber is evident in the churches, and timber bell-towers and spires are common. Throughout Sussex tiles for roofing are quite a feature, and also oak shingles for the timber spires and turrets.

Many of the Sussex churches are of the simplest design, and are built of the humblest materials. Some are on a very small scale, as the churches at Binsted, Burton, and Selham, while a few are really great, as those at Boxgrove, New Shoreham, and Winchelsea, not to mention the cathedral at Chichester.

There are various types of Sussex churches. First, we may notice those with only nave and chancel, and perhaps an aisle of later date; secondly, there are those with nave, chancel, and bell-turret; and thirdly there are the most numerous, those having nave, chancel, one or more aisles, and a western tower. A few instances occur

New Shoreham Church

of a cruciform building with a central tower, and there are three churches having round towers, those at Lewes, Piddinghoe, and Southease.   Besides the cathedral, the churches at Chiddingly, Northiam, Dallington, and East Preston have stone spires, and those at Heathfield, Winchelsea, and St Clement's, Hastings, have crypts.   There is little good stained glass in the churches; the fonts are

Worth Church

plain; and what is rather strange, considering that Sussex was once an iron-producing county, there is hardly any wrought-iron work.

From an antiquarian point of view the most interesting churches in Sussex are those at Worth, Sompting, and Bosham, which all date from the Saxon period.   Worth has the most complete ground-plan of a Saxon church which remains, and although it has been modernised, the chancel and arches are without doubt Saxon. Norman work may be seen in the churches at Old Shoreham, New Shoreham, Steyning, and Newhaven, and in the Cathedral. It is related that there were 150 churches in Sussex before the death of William I, but of course many of them have been rebuilt, still the number of small early Norman churches is peculiar to Sussex.

Towards the end of the twelfth century the round arches and heavy columns of Norman work began gradually to give place to the pointed arch and lighter style of the first period of Gothic architecture which we know as Early English, conspicuous for its long narrow windows, and leading in its turn by a transitional period into the highest development of Gothic—the Decorated period. This, in England, prevailed throughout the greater part of the fourteenth century, and was particularly characterised by its window tracery. The Perpendicular, which, as its name implies, is remarkable for the perpendicular arrangement of the tracery, and also for the flattened arches and the square arrangement of the mouldings over them, was the last of the Gothic styles. It developed gradually from the Decorated towards the end of the fourteenth century and was in use till about the middle of the sixteenth century.

The churches at Climping, Pevensey, and Horsted Keynes are representative of the Early English period; Winchelsea is a beautiful example of a Decorated church; and the Perpendicular is well shown in the two fine churches of St Clement and All Saints at Hastings.

Chichester Cathedral, in its present form, shows building mostly of the twelfth and thirteenth centuries. It was originally begun in 1091, and partly finished in 1108. It was burnt down in 1114, and another fire in 1186 hindered the progress of restoration. The work of rebuilding and remodelling went on from 1187 to 1244, and the Lady Chapel dates from 1304. The spire, originally erected in the fourteenth century, collapsed in 1861, and was rebuilt in 1866. The detached Bell Tower is a feature which now is peculiar to Chichester among all English Cathedrals.

The Reformation marks a distinct break in ecclesiastical architecture in Sussex. Before that change 70 religious houses, such as abbeys, priories, nunneries, and hospitals were dotted about the county. Many of these were fine specimens of the skill of the architect, but of only a few are there any remains to recall their former beauty. Battle Abbey was founded by William as a votive offering for his great victory at Hastings, and dedicated to St Martin. The Priory of Lewes was founded by William de Warenne and had possessions over almost all the kingdom. Both the heads of Battle Abbey and Lewes Priory were constantly summoned to Parliament, and their influence must have been very great. The remains of Battle Abbey are very interesting; those of Lewes Priory are but scanty. Of Bayham Abbey the cloisters and chapter-house remain, and there are also ruins of the religious houses at Boxgrove, Winchelsea, and Robertsbridge.

19. Architecture. (*b*) Military—Castles.

Few counties have such a numerous and interesting series of castles and defensive works as Sussex, and they are of every period and plan from the Roman castle of Pevensey to Camber Castle of the sixteenth century. Of course, the Norman period was the time when the largest

Camber Castle

number were built, and it is said that no less than 1100 were erected in various parts of the country during that time. However, we shall not be far wrong in stating that probably ten or twelve of the seventeen Sussex castles are of the Norman period. The Normans divided the county into the six "rapes" of Hastings, Pevensey, Lewes,

Bramber, Arundel, and Chichester, each having its own castle. This was a unique division of which we shall presently have to speak, and we find that each rape was given to one of William's chief followers.

The Norman castles were not all of the same size and importance, for while some were royal castles and meant for the defence of the country, others were built by the barons for the protection of their own territory and became the terror of the countryside. Before dealing with the Sussex castles in particular, it may be well to consider some general features connected with these buildings.

A castle of the best construction consisted of a lofty and very thick wall, with towers and bastions, enclosing several acres, and protected by a moat or ditch. Within this area were three principal divisions. First, there was the outer bailey, or courtyard, the approach to which was guarded by a towered gateway, with a drawbridge and portcullis. In this bailey were the stables, and a mount of command and of execution. Secondly, there was the inner bailey, or quadrangle, also defended by gateway and towers, within which stood the keep, the chapel, and the barracks. Thirdly, there was the donjon or keep, which was the real citadel, and always provided with a well.

Of all the Sussex castles, Pevensey stands first from its great historical interest. The present castle is mediaeval, and lies in the south-east corner of an enclosure, surrounded by a wall of Roman origin, which once bounded the famous town or fortress of Anderida. When the Saxons invaded Sussex, Anderida was blockaded and taken

by them, and for more than five centuries we hear little
of the town.   However, in 1066, it came into prominence
as the landing place of the Norman conqueror, and after
his victory at Hastings, he gave Pevensey to his half-
brother, Robert, who first built the castle within the
Roman walls.   The castle is divided from the rest of the
enclosure by a moat, and the main gateway, flanked by

Pevensey Castle

two towers, was reached by a drawbridge.   It is somewhat
difficult to distinguish the ruined fragments of this famous
keep, but the site of the chapel may be traced on the
turf, where also can be seen the plain old font, and the
opening of the well.

Hastings Castle, partly of Norman origin, is finely
placed on the cliff.   It is unapproachable on three sides,
and on the fourth a deep moat was dug for its defence.

It was probably built by William the Conqueror, and by him granted to Robert, Earl of Eu. The walls, now ruined, ran round three sides of the enclosure, and there are remains of some of the towers, arches, and windows.

Lewes Castle was given by William to William de Warenne. The ruins consist of two gateways and the keep, the latter of Edwardian origin. There was formerly a second keep, and the mound on which it stood is visible some distance away.

Bramber Castle was built by the Normans to defend the gap in the Downs made by the River Adur. There formerly stood here an old Saxon castle, and on its site, which was given by the Conqueror to William de Braose, a Norman castle was raised. The castle was besieged and taken by the Parliamentary army, which utterly destroyed it, so that only a fragment is left of this famous building.

Arundel Castle, together with the surrounding district, was given by King William to Roger de Montgomeri. It withstood three sieges, but at the last in 1644 Waller reduced it and laid most of it in ruins, so that of the old castle little remains but the Norman keep, which is very fine and commands an extensive view.

Like Arundel, Chichester was given to Roger de Montgomeri, who built a castle here which has entirely disappeared. The castle was in the north-east quarter of the city, and the city walls were not strong enough to enable it to hold out for more than ten days when Waller besieged it in 1643.

Hurstmonceux Castle is of brick, and one of the earliest buildings erected of this material since Roman times. It lies low, and is rather a fortified mansion than a castle. In plan it is almost a square with octagonal angle towers, and four others, one in the centre of each side. The main gateway is the most beautiful part of the castle. The two flanking towers are capped by smaller watch-towers, and joined by an archway. The ruins are extensive and picturesque, and comprise the great hall, two dungeons, the small chapel, the postern gate, the kitchen and offices.

Bodiam Castle

Bodiam Castle is a beautiful and interesting ruin. It was built in a low situation in the valley of the Rother, and has a well-filled moat. The surrounding walls of the enclosure are very solid, and the castle is rectangular in plan. There is a round tower at each angle, and in the middle of each side a square tower. The chapel, the great hall, the buttery, and the kitchen are the chief remaining portions.

The Ypres Tower at Rye, which stands on the cliff, was built in Stephen's reign by William of Ypres, and was for some time the only defence of the town. It is square in plan, with round towers at the angles, and was long used as a prison.

Camber Castle lies on the low coast between Rye and Winchelsea. It is the most modern of the Sussex castles, and was built by Henry VIII in 1531 for the defence of the coast. It is not of much interest, and its round central keep, exterior wall, and towers at regular intervals, only remind us of the fact that by the time of the Tudors the days of castles had passed away.

20.   Architecture.   (c) Domestic— Famous Seats, Manor Houses, Cottages.

After the Wars of the Roses there was no longer necessity for castles and fortified houses, and in Tudor times the houses of the great nobles were built less like fortresses and more as comfortable homes for the owner, his family, and his servants. Besides these larger mansions there were also many good manor-houses, where the lord of the manor lived in the heart of his own domain. Sussex has no great houses of such architectural splendour, or with such historical associations, as Hatfield House in Hertfordshire, or Penshurst Place in Kent, but it has some really good houses of a less pretentious character, dating from the fifteenth and sixteenth centuries and even earlier.

The architecture of a county is always influenced by the building materials that are found within its borders, and the remarks on this subject in Chapter 18 apply also to the domestic buildings. Owing to the ease with which timber could be obtained from the Sussex forests, we naturally expect to find that wood is one of the chief materials used in the construction of Sussex houses and cottages. For some of the larger houses stone is used, both from Sussex quarries and from Caen in France. For the smaller houses and cottages, oak timber and clay plaster, or "wattle and daub," were common. In some districts flints, chalk, and sandstone were freely used for the walls, and Horsham slabs for the roofs. Here and there may be seen farm-houses and cottages with coverings of reed-thatch ; while in most parts tiled roofs are quite a feature of the architecture. It should be noted that to the end of the sixteenth century little window glass was used; and that bricks, so much used by the Romans, were not reintroduced till about the fifteenth century.

Let us now consider a few of the famous old houses of Sussex. We cannot do better than begin with Crowhurst manor-house, near Hastings, dating from the thirteenth century. It was built in the form of a parallelogram with a small porch, and contained only three rooms, besides a hall. Although scanty, the ruins are picturesque, and the east window of the large room has good mouldings.

There are very early manor-houses at Malling near Lewes, at Preston, Portslade, Ferring, and Goring, and some remains of a thirteenth-century court-house at Winchelsea. But the ruins of Mayfield Palace, once the country house of the archbishops of Canterbury, from Dunstan to Cranmer, are the most important of their class in England. After lying in ruins for a very long time, the place was bought by the Roman Catholics and

The Great Hall, Mayfield

turned into a convent. The Great Hall with its remarkable roof and fine windows is the chief feature, but this is now used as a chapel.

When the Tudor period is reached the remains of the fine domestic buildings are numerous. Brede Place, built of brick, stone, and half timber, is very picturesque, and retains the chapel and many of its original fittings. Cowdray House, near Midhurst, is a grand example of a Tudor mansion. The ivy-clad remains of this house, half castle and half manor-house, give one an excellent idea of the kind of building that was common at this period—the early sixteenth century. The entrance gateway is square with turrets at the angles; the hall has some elaborate ornaments; and one of the chief features, the six-transomed oriel window, is particularly fine.

Brede Place

Many good Sussex mansions date from the sixteenth and seventeenth centuries, but we can only mention those at Cuckfield, Parham, Wiston, and Glynde. The farm-houses and buildings of West Sussex are worth a passing notice. They are generally picturesque, having roofs either of thatch or tiles; the barns are weather-boarded.

There are few English counties that can boast of such pretty cottages as Sussex, and charming examples of them may be seen in many of the rural districts, particularly in the neighbourhood of Henfield. They are often built of dark red bricks, with tiles of various sizes and shapes as a facing to the gables, as well as a covering to the roofs. Some of the cottages are built partly of wood, but in

Cowdray House, Midhurst

nearly all cases they are well adapted to their situation. The chimneys are one of the characteristics of these Sussex cottages, and while not so ornamented as those in Kent, they are larger and more massive.

The Sussex cottages of this picturesque type are passing away, and are being replaced by flimsy buildings, devoid of taste and individuality. The old cottages grew in beauty with age, but the modern structures of yellow bricks and blue slate roofs will always be an eyesore, and certainly no one will ever admire their rectangular windows or

Old Houses, Petworth

their miserable chimney-pots. An artist delights in sketching the fine old half-timbered houses and cottages at Rye,

Winchelsea, Horsham, and Petworth, but who would ever think of putting on canvas the buildings that are springing up in some of the modern towns ! The old and charming Sussex cottages were suited to the wants of the occupiers, and yet seemed to be inspired by the genius of the place.

21. Communications—Past and Present —Roads, Railways, Canals.

It has been well said that of all the marks made by the Romans in this island, the most distinct and ineffaceable was that left by them as road-makers. Often indeed their works survive only as boundaries between parishes or counties, but sometimes we can see the track still going straight to its mark over hill and dale, and we say instinctively, "That must be a Roman road." There is no doubt that the Romans, in some cases, made use of the British trackways, which were improved and increased ; but throughout our country the Romans were the great road-makers, and, strange to say, their chief highways are known to us for the most part by the names given to them by our Anglo-Saxon forefathers.

If we look at the map of Sussex we see that the chief roads run from south to north, and the reason for this is obvious. The main roads of Sussex converge on London, the chief seat of the kingdom. The Romans carried their roads from openings on the sea-coast, such as the mouth of the Arun, the mouth of the Ouse, or from Pevensey, through gaps in the South Downs and the North Downs onwards to the metropolis.

It will not be possible to deal fully with the Roman roads, but we may just touch upon the one called Stane Street, or Stone Street, connecting London with Chichester. This latter town, which was built by the Romans and named *Regnum*, was laid out in a geometrical way, and its four main streets run in north, south, east and west directions. The east street was Stone Street, and

Chichester Cathedral

was probably the old British trackway which the Romans paved with stone. From Chichester, Stone Street runs direct to London.

There is every reason to believe that the main roads of Sussex continued to be fairly good through the Middle Ages, but by the fifteenth and sixteenth centuries they were entirely worn out. Many of the newer roads were mere country lanes, impassable except by pack-horses. In fact, the pack-horse was at this time almost the only means of transport as far as trade was concerned. Even as late as the beginning of the nineteenth century, Sussex was considered one of the worst counties in which to travel, and many stories are told of royal and other passengers who got upset, or stuck in the mud.

Especially bad were the roads in the Weald, and reference has already been made to the fact that Prince George of Denmark when visiting Petworth, in 1703, took six hours to travel nine miles. Defoe in his *Tour*, published in 1724, tells us that he saw a lady drawn to church by a team of oxen, as the road was impassable for horses. At a later period of the eighteenth century, Horace Walpole came to difficulties in travelling to Robertsbridge. He says, "The roads grew bad beyond all badness, the night dark beyond all darkness, our guide frightened beyond all frightfulness. However, without being at all killed, we got up, or down—I forget which, it was so dark—a famous precipice called Silver Hill, and about ten at night arrived at a wretched village called Rotherbridge. We had still six miles hither, but determined to stop, as it would be a pity to break our necks before we had seen all we had intended."

Horace Walpole further relates that, in those parts of Sussex he visited, it was the custom for young gentlemen to drive their curricles with a pair of oxen. Now, however, oxen are rarely seen on the Sussex roads, but on the hill-sides a few of the farmers still plough with them. The black oxen of the hills are of Welsh stock, and the

" kews," as their shoes are called, may still be seen on the walls of a smithy here and there. Shoeing oxen is a difficult task, since to protect the smith from their horns, they have to be thrown down, their necks held in a fork, and their feet tied together.

Arthur Young, who travelled through Sussex at the end of the eighteenth century, reports that " the turnpike roads in Sussex are generally well enough made. The cross roads upon the coast are mostly kept in good order ; but in the Weald the cross roads are in all probability the very worst that could be met with in any part of the island." In the early years of the nineteenth century a great change took place in the condition of the Sussex roads. " Clinkers," as the Sussex ironstones were called, were used to mend the more important roads with good results. New roads were also made, and now it is only possible to complain of the country lanes.

The road from London to Chichester enters the county from Haslemere in Surrey ; the road from London to Rye and Winchelsea runs for upwards of twenty miles along the borders of Sussex and Kent ; and the road to Hastings branches from it at Flimwell, and thence proceeds through Robertsbridge and Battle to the sea. But of all the Sussex roads, the most nearly perfect, and certainly the most fashionable of all, is the Brighton Road. This is not one road only, but three roads, and in point of fact, according to some authorities, there are five roads. In the coaching days, which were at their height during the Regency and the reign of George IV, there was no town in England that could be reached by

Lewes Castle: the Entrance Gate

so many different routes as Brighton. Of these the favourite was the New Road, which went by Croydon, Merstham, Redhill, Balcombe, and Cuckfield, and made the distance about $51\frac{1}{2}$ miles. The longest and oldest route was through Croydon, Uckfield, and Lewes, a distance of $58\frac{1}{4}$ miles. It is related that no fewer than eighteen coaches left London for Brighton each day in 1821, and that the distance was covered by the fastest in five hours and a quarter.

We will now leave the roads, and turn our attention to the canals and railways of the county. Of canals there are several, but at present they are somewhat neglected. The most important is in connection with the Arun, which is joined to Portsmouth and Chichester by the Arundel and Portsmouth Canal. The West Rosher Canal joins Petworth and Midhurst with the Arun ; and the Arun and Wey Canal, which was completed nearly one hundred years ago to connect these two rivers, is now practically closed.

With regard to railway communication, Sussex is almost entirely served by the London, Brighton, and South Coast Railway. The line to Brighton was opened in 1841, and within the next few years branches were made to Chichester on the west, and to Hastings and Eastbourne on the east. The South-Eastern and Chatham Railway has a line from Tunbridge Wells to Hastings and to Rye, and the London and South-Western Railway a branch from Petersfield to Midhurst.

22. Administration and Divisions— Ancient and Modern.

In an earlier part of this book we found that Sussex may be accepted as the best typical instance of an old English kingdom becoming a county, and as the modern representative of an old independent Teutonic commonwealth. We are now in a position to consider how its present local independence, or, in other words, its present administration of county affairs, is to a large extent the result of its past history. Let us therefore, before dealing with the present forms of administration, glance at the ancient forms of government. It must always be clearly understood that, though many changes have been introduced into our parochial and county government, it has ever been the care of most of our rulers to graft, as it were, new ideas on to the old English institutions. Thus it comes about that what we often consider modern methods of government may be traced back to Saxon days, more than a thousand years ago.

In early English times the government of a county or shire was partly central, from the county town, and partly local, from the hundred or parish. The chief court of Sussex was, in the earliest times, the Shire-moot, which met twice a year. Its chief officers were the Ealdorman and the Sheriff, the last of whom was appointed by the King. Each county was divided, in Saxon times, into Hundreds, or Lathes, or Wapentakes. Sussex was divided into hundreds, and it is probable that, at first, each of

these divisions contained one hundred free families. Each
hundred had its own court, the Hundred-moot, which
met every month for business.

Although probably dating not from Saxon, but from
Norman times, there is another division of the county
which is quite peculiar to Sussex. The various hundreds
were grouped into six divisions, or Rapes, and each portion
extends from the northern border of the county to the
sea. It is also worth noting that besides possessing a sea-
frontage, each Rape has within it one castle, or other
important station for defence and protection. It must,
however, be clearly understood that the Rape was only a
geographical division, and had no administrative or judicial
position.

Now let us return to the other, and earlier Saxon
divisions. Each hundred was divided into townships, or
parishes, as they are now called. Each township had its
own *gemot*, or assembly, where every freeman could
appear. This *gemot*, or town-moot, made laws for the
township, and appointed officers to enforce these *by*-laws,
or laws of the town. The officers of the town-moot
were the reeve and the tithing-man, the last of whom
corresponds to our policeman. The reeve presided over
the township court, which was held whenever necessary.

Besides these courts of the shire, the hundred, and the
township, there were also courts of the manor, as the
separate holdings of land were called. The manors were
of different sizes, sometimes they were as large as the
township, and sometimes they were parts of the township.
The manors were held by their owners, or lords of the

manor, as they were called, on various conditions. For example, they had to render service or homage to the King, and were allowed to sub-let their manors. The manor-courts were of various names, such as court-leet, court-baron, and customary court. In these courts, the lord and his tenants met, and settled the affairs belonging to the manor, such as those relating to the common fields, the rights of enclosure, and the holding of fairs and markets. These manor-courts are still held in many parts, but they have long since lost the importance they once had, and we only refer to them here to show how interesting it is to remember that they are survivals dating back for more than one thousand years.

Having briefly considered the ancient forms of administration in the county, we are now enabled to form a better idea of the present mode of administering local affairs. The chief county officers at present are the Lord Lieutenant, who owes his appointment to the Crown, and is generally a nobleman or a large landowner, and the High Sheriff, who is chosen every year "on the morrow of St Martin's Day," November 12.

Sussex is divided into two portions, known as East Sussex and West Sussex, each of which has its own County Council, a mode of government which was introduced in 1888, and which corresponds to the ancient Shire-moot. The County Council of East Sussex has 17 Aldermen and 52 Councillors, and the County Council of West Sussex has 16 Aldermen and 49 Councillors. The County Councillors are elected every three years to their position, while the Aldermen are co-opted by the

Councillors for a term of years. Lewes is the centre of county business for East Sussex, and for West Sussex, meetings are held at Chichester and Horsham.

For local government in towns and parishes, an Act was passed in 1894, when new names were given to the governing bodies formerly known as vestries, local boards, highway boards, etc. In the large parishes, the chief

Battle Abbey: the Cloister Front

governing bodies are now called District Councils, and of these there are 15 in Sussex, while the smaller parishes have Parish Councils, or Parish Meetings. But whether District Councils or Parish Councils, they represent the old town-moots, and the members, chosen by the people, are elected to manage the affairs of the district, and generally to advance its interests.

There are some towns in Sussex that have different and larger powers of government than the parishes. These are called Boroughs and are as follows :—Brighton, Hastings, Rye, Chichester, Eastbourne, Hove, Arundel, Worthing, and Lewes. Of these nine Boroughs, the first two are called County Boroughs, having the powers of a County Council.

Sussex has also 21 Poor Law Unions, each having a Board of Guardians, whose duty it is to manage the workhouses, and appoint various officers to carry on the work of caring for the poor and the aged.

For the administration of justice, East Sussex and West Sussex have their own police force. The Quarter Sessions are held at Lewes, Chichester, Petworth, and Horsham, and there are 18 Petty Sessional Divisions, each having its own magistrates, or justices of the peace.

In the earliest period of our history, the Church existed as a body before the State, and its mode of government is much the same to-day as it was in those far-off days. In Saxon times, our country was divided into dioceses, or sees, over which were placed bishops. The northern dioceses and bishops were under the rule of the Archbishop of York, and the southern dioceses and bishops under the Archbishop of Canterbury. Sussex was in the diocese of Selsey, but after a time the see was removed to Chichester, where it has since remained. Practically the whole of Sussex is in the diocese of Chichester, but there are a few parishes in the dioceses of Canterbury, Winchester, and Rochester. At one time the ecclesiastical parish coincided with the civil parish, but now while there are 338 civil parishes, there are 377 ecclesiastical parishes in Sussex.

Matters relating to education are managed by Education Committees in Brighton, Hastings, Bexhill, Eastbourne, Hove, Lewes, Chichester, and Worthing ; but for all the other parishes, the County Councils of East Sussex and

Christ's Hospital, Horsham

of West Sussex have appointed Education Committees to control Secondary and Elementary Education.

For parliamentary representation, the county is divided into the following six divisions :—Horsham, Chichester, East Grinstead, Lewes, Eastbourne, and Rye, each electing one member. The borough of Brighton elects two members, and Hastings one member. It is interesting to note that the Reform Act of 1832 disfranchised the following towns :—Bramber, East Grinstead, Steyning, Winchelsea, Arundel, Horsham, Midhurst, and Rye.

23. The Roll of Honour of the County.

In bringing our survey of Sussex to a close it will be of interest in this chapter to associate the names of the famous men of the county with the places where they were born, or resided, or with which they were in some way connected. Certainly we shall be all the more attached to our county if we know a little of the men who have shed lustre upon it and have helped to make it famous. The love of our own parish is quickened if we realise that some man of note has lived in it, and this spirit of local pride is the beginning of patriotism for our native land, and for the Empire as a whole.

Among the many persons of royal birth who have been associated with Sussex, we can mention but a few. King Alfred the Great lived at Dean, where he entertained Asser, who afterwards became his biographer. King Alfred had much property in the county, and in his Will many Sussex names of places occur. Harold, the last of the Saxon Kings, had extensive possessions in the county, and when he left England on his excursion to Normandy, he sailed from the Sussex port of Bosham, a place which was then of much importance. The name of William the Conqueror will always be associated with the defeat of Harold at Hastings, and also with the division of the county among some of his Norman barons. There is not space to tell of the other English monarchs who have left memories in Sussex, but mention must be made of George IV, who as Prince of Wales so often visited Brighton, built the Pavilion, and established the popularity of the town as a seaside resort.

When we consider the divines we find that Sussex has a long list of them associated with various places.

Parsonage Hall, West Tarring

Archbishops, Bishops, and Deans besides others of lesser note are connected with one town or another. Dunstan, in the tenth century, lived at Mayfield, where are still ruins of the palace of the archbishops. Archbishop Becket in the twelfth century had a house at West Tarring, and Archbishop Juxon, who attended King Charles I on the scaffold, was born at Chichester. Chillingworth, a famous religious writer, was taken prisoner at the siege of Arundel, but was allowed to go to the bishop's palace at Chichester, where he died in 1643, and was buried in the cathedral of that city. Cardinal Manning was at one time Archdeacon of Chichester; and Bishop Wilberforce lived for a long time at West Lavington. There he was buried, and the church has been restored in memory of him. Dr Hook, author of *The Lives of the Archbishops of Canterbury*, was Dean of Chichester; and Bishop Hannington, who was murdered in Africa in 1885, was both a native of and a curate at Hurstpierpoint. Augustus and Julius Hare, the joint authors of *Guesses at Truth*, were connected with Hurstmonceux, and Julius was Archdeacon of Lewes.

The names of Wilberforce and Cobden occur among the Sussex worthies as statesmen. The first, William Wilberforce, was M.P. for Bramber, and he will ever be remembered as the liberator of the slaves; the second, Richard Cobden, was born at Midhurst, and educated at the grammar school of that little town. His work in connection with Free Trade is well known, and at the close of his career he was buried in the churchyard of West Lavington in 1865.

The only man of action of great note who lived in the county was General Eliott. He was famous as the defender of Gibraltar, at the great siege from 1779–1782. For his success he was created Lord Heathfield, a title derived from the fact that his residence was at Heathfield Park, in Sussex.

When we turn to historians, we find that Gibbon,

the author of *The Decline and Fall of the Roman Empire*, spent much of his time at Sheffield Park, and was buried at Fletching in 1794. Mark Antony Lower, the author of the *Worthies of Sussex* and the *History of Sussex*, was a native of the county and a schoolmaster at Lewes.

Edward Gibbon

There are many men of letters who are associated with Sussex. First must come the celebrated Selden, who was the "great dictator of learning of the English nation" during the Civil War. He was born at Salvington, educated at Chichester, and sat in several Parliaments. He was one of those appointed to draw up the Petition of Rights, and gained much fame as a politician and jurist. Horace Walpole visited Hurstmonceux, and as man of letters, novelist, and critic takes high rank.   His *Letters*

Field Place, Warnham
(*Shelley's Birthplace*)

help us to understand the social life of the age in which he lived.   Thackeray, the novelist, knew Brighton and Winchelsea well, and the latter place figures in his story of *Denis Duval*, while the former he called " Dr Brighton," on account of its health-giving powers.   Thackeray's words are " One of the best physicians our city has ever known is kind, cheerful, merry Doctor Brighton." William Black, another Victorian novelist, lived at Brighton, and *That Beautiful Wretch* and *Kilmeny* owe much to his knowledge of this town. Black was buried at Rottingdean.

Of poets connected with Sussex we have many from Pope onwards. It is said that Pope wrote *The Rape of the Lock* at West Grinstead. Collins, a poet now little read, was born at Chichester in 1721, and died near the Cathedral cloisters in 1759. Byron lived at Hastings and Littlehampton; and Shelley was born at Field Place, Warnham, in 1792. There are also memories of Cowper at Eartham, and of Campbell and Coventry Patmore at Hastings.

Among artists, we may mention Sir Edward Burne-Jones, one of the most famous of the Pre-Raphaelites, who lies buried at Rottingdean, in which church is some stained glass designed by him.

We will bring this list of Sussex worthies to a close by mentioning that Sir Charles Lyell, the eminent geologist, was educated at Midhurst Grammar School, and that Herbert Spencer, who ranks high as a philosopher, lived at Brighton.

24. THE CHIEF TOWNS AND VILLAGES OF SUSSEX.

(The figures in brackets after each name give the population in 1901, and those at the end of each section are references to the pages in the text.)

Aldrington (6840) is a coast parish three miles west of Brighton. It has a canal and small floating basin.

Alfriston (534) is a large and picturesque village at the entrance of the gap by which the Cuckmere makes its way through the Downs. It has a fine church, which has been called the "Cathedral of the Downs," some ancient houses, and a market-cross.

Ardingley (1346) on high ground overlooking the Ouse, is five miles south-west of Horsted Keynes. Its church is of considerable interest, and *Wakehurst Place* is a good example of an Elizabethan mansion.

Arundel (2739) is a municipal borough, five miles above the mouth of the Arun. Its fine baronial castle is picturesquely situated, and has long been associated with the Howard family. Of the old castle little remains except the keep, but the present Duke of Norfolk has built a new castle, and this is the chief feature in the town. The parish church is interesting, and there are some old houses of considerable merit. (pp. 17, 19, 31, 32, 57, 58, 65, 83, 84, 95, 101, 103, 121, 123.)

Ashburnham (570), about four miles west of Battle, is noteworthy as the last place in Sussex where iron-works were carried on. *Ashburnham Place* is a fine mansion surrounded by a large, well-timbered park. (p. 65.)

Battle (2996) is a market-town seven miles north-west of Hastings. The Battle of Hastings was fought here in 1066, and there are the remains of the Abbey founded by William the Conqueror. Of the ruins, the most interesting portion is the gateway. (pp. 58, 59, 81, 88, 99.)

Bexhill-on-Sea (12,213), five miles west of Hastings, is a rising sea-side resort. (pp. 88, 122.)

Bignor (104), six miles north of Arundel, has the interesting remains of a Roman villa, which was discovered in 1811. The principal remains are three fine tessellated pavements. (p. 60.)

Billingshurst (1591) is a very old parish, seven miles south-west of Horsham.

Bodiam (252) on the River Rother, three miles east of Robertsbridge, has some beautiful ruins of a castle. (p. 104.)

Bognor (6180) is a watering-place midway between Selsey Bill and the mouth of the Arun. The sands are firm and extensive, but the country inland is not interesting. (pp. 6, 40, 69, 88.)

Bosham (1149) is a small fishing village about four miles from Chichester. Many traces of Roman occupation have been found. Earl Godwine lived here, and it was from Bosham that Harold sailed on his journey to Normandy. The parish church is of great antiquity, having both Saxon and Norman work. (pp. 57, 71, 97.)

Brighton (123,478) is a parliamentary and county borough, and a large and fashionable watering-place. It was formerly known as *Brighthelmstone*, and for some centuries before 1780 was only a fishing village. It was patronised by George,

Prince of Wales, afterwards George IV, and has ever since been the great pleasure resort of Londoners in Sussex. Its principal attraction is the magnificent sea-front extending for upwards of four miles. Its Pavilion, built by George IV, is a curious pile of domes and minarets, and now belongs to the town, being used as a place of entertainment. The Aquarium is perhaps the best in England, and there are numerous other buildings of interest. There are two fine promenade piers; and the herring and mackerel fisheries are of importance. The Devil's Dyke is in the neighbourhood, and the view over the Weald from its summit is extensive and beautiful. "London-by-the-Sea" has been visited by countless celebrities, and Thackeray described it as "Dr Brighton," on account of its health-giving properties. (pp. 1, 6, 13, 15, 16, 19, 26, 28, 36, 37, 41, 52, 58, 68, 69, 85, 88, 121, 122, 124, 127, 128.)

Broadwater (1187), a little north of Worthing, has a beautiful church, with a massive tower. Parts of the building are of great antiquity, and of the richest style of architecture.

Burwash (1977), eight miles north-west of Battle, is on high ground in the valley of the Rother. The church is of interest, and is certainly of Norman, and probably of Saxon date. (p. 65.)

Buxted (2038) was formerly one of the great iron-towns of the Weald, of which it was the centre. The first English cannon was cast here in 1543. (pp. 64, 65.)

Chichester (12,244), a borough and city near the head of Chichester Harbour, is of great antiquity. In the days of the Romans it was called *Regnum*, and was the headquarters of Vespasian. Its Roman origin is traced in the four nearly straight streets, answering to the points of the compass, and meeting at the handsome market-cross, which was completed about 1500. The city takes its name from Cissa, a Saxon chief, who conquered

9—2

it in 477. The ancient walls still mark the site of the old city. The Cathedral, with five aisles, is very interesting and dates from the eleventh century. On the north-west side of the cathedral is the Bell Tower, or campanile, which is the only English example of a detached belfry adjoining a cathedral. Chichester is the headquarters of the West Sussex County Council, and has extensive corn and cattle markets. (pp. 3, 12, 39, 45, 55, 58, 60, 65, 76, 77, 78, 84, 88, 89, 91, 93, 95, 99, 101, 103, 112, 120, 121, 122, 128.)

Cuckfield (1813), two miles west of Haywards Heath, is a place of some importance, having beautiful views over the Downs. The church is Early English with some good monuments. (pp. 19, 108, 116.)

Ditchling (1253) has a fine church, and some good timber-built houses in the village. Ditchling Beacon, one of the highest points of the Downs, has traces of some ancient camps, perhaps of Roman date. (pp. 15, 62.)

Eastbourne (43,344) is a municipal borough and fashionable watering-place. It has more than doubled its population since 1881, and has many attractive features. It is situated close under Beachy Head and is backed by the Downs. The streets are broad and lined with trees; the houses are well built of stone; and the parade is nearly three miles long. Devonshire Park is one of the chief attractions of the place, and there are several fine institutions. The parish church in the old town is ancient, and of great interest. The churchyard used to be surrounded by a moat. (pp. 26, 36, 37, 43, 52, 57, 61, 69, 71, 85, 87, 88, 91, 121, 122.)

Fletching (1088), eight miles north of Lewes, stands on the Ouse in the midst of a well-wooded district. Gibbon, the historian, often stayed at Sheffield Park and the fine church contains his tomb. (p. 126.)

Frant (1692), three miles south of Tunbridge Wells, is a delightful village on a hill over 600 feet high. Eridge Park, in the neighbourhood, is charming and has fine and extensive views.

Grinstead, East (6094) is a market-town pleasantly situated on a hill with a fine view over the Forest Ridge. In the High Street are some quaint timber-built houses, and the church has a fine tower. *Sackville College* is an old almshouse dating from 1609. (pp. 122, 123.)

Grinstead, West (1503) is famous for its beautiful park and its deer. The church with its shingled spire has some Norman work. (p. 128.)

Hailsham (4197) is a thriving little town seven miles north of Eastbourne, with one of the largest cattle and sheep markets in Sussex. (p. 58.)

Hastings (65,528), one of the Cinque Ports, is a popular summer resort in East Sussex. St Leonards is joined to Hastings by a row of terraces and parades. The ruins of the castle are finely situated and the old town beneath the Castle Hill is very picturesque with its red-tiled roofs. The harbour lies at the east end of the town, and the fishery is of some importance. (pp. 1, 6, 36, 37, 43, 52, 56, 59, 62, 69, 72, 74, 78, 88, 91, 97, 98, 101, 102, 121, 128.)

Haywards Heath (3717) is a small market-town, and the largest cattle sale in Sussex is held here.

Heathfield (3745) has a good church with a noteworthy tower and spire. *Heathfield Park* is well-wooded and was once the residence of General Eliott, the defender of Gibraltar. (pp. 19, 54, 83, 97, 125.)

Horsham (9446) is an old and picturesque market-town on a branch of the Arun. It has some trade in corn and timber, and the industries include malting and brewing. The church, dating from the time of King John, is noteworthy. *Christ's Hospital*, known as the Bluecoat School, has been established at Horsham in a fine pile of buildings. (pp. 7, 15, 18, 58, 59, 61, 110, 120, 122, 123.)

Hove (29,695) is continuous with Brighton, and contains many handsome streets and public buildings. It is a fashionable resort, and has the Sussex County Cricket Ground. (pp. 41, 88, 121, 122.)

Hurstmonceux (1429) has the remains of a red brick castle, perhaps the most picturesque ruin in Sussex. (pp. 57, 62, 103, 125, 127.)

Hurstpierpoint (3033) is a market-town eight miles north of Brighton. (p. 125.)

Icklesham (1447), about one and a half miles west of Winchelsea, has a beautiful church with a Norman tower.

Lancing (1244) is a sea-bathing resort, about two miles from Shoreham. Besides the Norman church *Lancing College* on the hill slope is the chief building. (p. 58.)

Lewes (11,249), a borough and market-town on the Ouse, is the county town of Sussex. It is noteworthy as being the scene of the battle (1264) between Henry III and Simon de Montfort. There are ruins of a castle and of a priory, and some of the old houses are interesting. Lewes is an important agricultural centre with an annual sheep-fair. (pp. 17, 26, 58, 59, 62, 83, 88, 91, 97, 99, 101, 103, 121, 122, 126.)

Littlehampton (5950) is a few miles to the south of Arundel of which town it is the port. The town is much frequented as a watering-place, and the sands are very extensive. (pp. 19, 57, 62, 128.)

Mayfield (3164), eight miles south of Tunbridge Wells, is an interesting little town near the source of the Rother. The

Petworth Church

remains of the palace of the archbishops of Canterbury are note-worthy. Mayfield was formerly one of the centres of the Sussex iron trade. (pp. 59, 65, 107.)

Midhurst (1650) is a little market-town on the Rother. At its Grammar School Sir Charles Lyell, Richard Cobden, and other famous men were educated. (pp. 7, 85, 108, 116, 123, 125, 128.)

Newhaven (6373) is a seaport town at the mouth of the Ouse. It is the chief place of embarkation for Dieppe, and steamers also go to Honfleur, Caen, and the Channel Islands. It has also considerable trade in corn, coal, and timber. (pp. 19, 42, 44, 58, 69, 72, 98.)

Northiam (1024), eight miles north-west of Rye, has a church with a remarkable Norman tower.

Petworth (2503) is a little market-town on the west Rother. Its streets are very narrow but there are many good houses. *Petworth House* is one of the finest mansions in the county, and has a rich collection of paintings and statuary. (pp. 14, 31, 60, 110, 113, 116.)

Pevensey (468), on the River Ashburn, is a very ancient place, and occupies the site of the Roman fortress, *Anderida*, of which the walls remain. The castle, now a picturesque ruin, was built soon after the Conquest. (pp. 6, 8, 17, 43, 45, 61, 67, 71, 73, 76, 83, 87, 89, 98, 101.)

Portslade-by-Sea (5217) is a modern and well-frequented seaside resort on the east side of Shoreham Harbour. The old village of Portslade is a short distance inland.

Pulborough (1725), on the Arun, is an ancient town with a large and remarkable church. Many Roman remains have been found in the neighbourhood, and the great *Stane Street* passed through the village. (pp. 61, 62, 95.)

Rotherfield (6462), near the source of the Rother, stands about 500 feet above the sea. A Saxon church was founded here in the eighth century, but the present building is of later date. (p. 19.)

Rottingdean (1992), a coast parish and sea-bathing resort, is about four miles east of Brighton. The little village is very picturesque, and in the church are buried Sir Edward Burne-Jones, the painter, and William Black, the novelist. (pp. 42, 128.)

Rye (3990), an ancient town and Cinque Port, is at the mouth of the Rother. It was once a famous seaport, but owing to changes in the coast-line the town is now two miles from the sea. Rye is quaint and picturesque, and besides a large and interesting parish church has a remarkable building, long used as a prison, called the *Ypres Tower*. Rye has now some trade in corn, coal, timber, wool, and hops. (pp. 6, 11, 17, 19, 47, 48, 56, 57, 58, 68, 69, 72, 74, 84, 105, 110, 121, 122, 123.)

Seaford (2615), at the mouth of the Ouse, was once a borough sending two members to Parliament, and also a member of the Cinque Ports. The harbour is now completely silted up, but the town is of some importance as a seaside resort. (pp. 73, 75.)

Selsey (1258), seven miles south of Chichester, was in Saxon times a place of great importance, and the seat of a bishopric. A great part of the old town and the site of the cathedral were washed away, and the bishop's see was removed to Chichester. The people are mainly engaged in the fisheries, which are of some value. (pp. 13, 27, 40, 69, 88, 93.)

Shoreham, New (3837) is a seaport at the mouth of the Adur, which is here crossed by a suspension bridge. Its church is one of the finest in Sussex, and dates from early Norman times. New Shoreham has trade with France, Holland, and the north of Europe, and there are industries connected with shipbuilding and the fisheries. (pp. 16, 19, 41, 56, 57, 69, 72, 95, 98.)

Shoreham, Old (281), one mile north of New Shoreham, is noted for its fine cruciform church, with a central tower, and built of flint. (pp. 56, 98.)

Steyning (1752) is a market-town, five miles north-west of Shoreham. The place is of great antiquity, being mentioned in King Alfred's Will. The church is interesting and of Norman date. Steyning formerly returned two members to Parliament. (pp. 19, 91, 123.)

Farm house, Warnham

Wadhurst (3232), six miles south-east of Tunbridge Wells, has sandstone quarries. Its church is remarkable for its pavement of about 30 iron grave-stones. (p. 60.)

Winchelsea (157) is a very ancient town, three miles west of Rye. In early times it was a port of great importance, but in the thirteenth century it was destroyed by inundations of the sea. Edward I rebuilt it, but owing to the retreat of the sea, it declined, and is now only a village with a parish church.   (pp. 37, 43, 47, 55, 68, 69, 72, 74, 95, 97, 98, 99, 107, 110, 123, 127.)

Worth (4297) is famous for its church with a Saxon ground-plan which makes it one of the most interesting in England. (p. 97.)

Worthing (16,996), the largest town in west Sussex, has risen from a poor fishing-village to a considerable watering-place. The climate is mild, and in the neighbourhood are nurseries and glasshouses devoted to the cultivation of tomatoes, grapes, etc. for the London markets.   (pp. 62, 69, 121, 122.)

Fig. 1.   Diagram showing the area of Sussex (1459 sq. miles)
          compared with that of England and Wales

Fig. 2.   Diagram showing the population of Sussex (605,202)
          compared with that of England and Wales

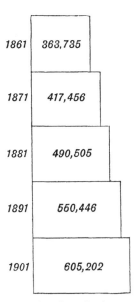

Fig. 3. Diagram showing the increase of Sussex
population from 1861 to 1901

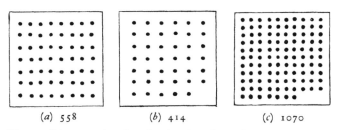

Fig. 4. Diagram showing the density of population to a square
mile in (a) England and Wales, (b) Sussex, and (c) Lanca-
shire. Each dot represents 10 people.

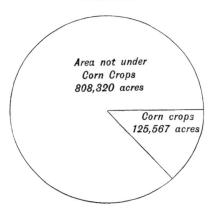

Fig. 5.   Diagram showing the area under
Corn Crops in 1905

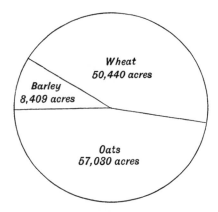

Fig. 6.   Diagram showing the proportionate areas
growing Wheat, Barley, and Oats in 1905

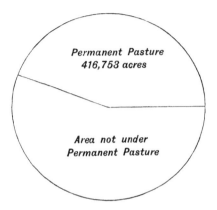

Fig. 7.   Diagram showing the area under
Permanent Pasture in 1905

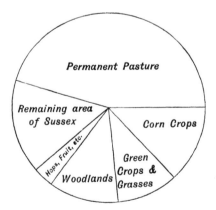

Fig. 8.   Diagram showing the proportionate acreage
under Crops, Grass, etc., in 1905

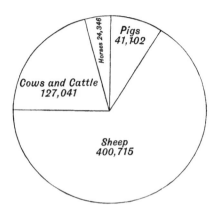

Fig. 9. Diagram showing the proportionate number of
Sheep, Cows and Cattle, Pigs and Horses in Sussex
in 1905

www.ingramcontent.com/pod-product-compliance
Ingram Content Group UK Ltd.
Pitfield, Milton Keynes, MK11 3LW, UK
UKHW042145280225
455719UK00001B/125